Thermal Plasmonics and Metamaterials for a Low-Carbon Society

In this edited volume for researchers and students, experts in thermal plasmonics and metamaterials technologies introduce cutting-edge energy and resource conservation techniques and environmentally friendly solutions in areas including energy generation and harvesting and radiative cooling.

Through this book, readers will gain an in-depth understanding of the metamaterials and thermal plasmonics technologies used for such devices and the real-world applications of these technologies. This book is divided into three broad sections to address different aspects of these devices. The first section presents research on materials that can control thermal radiation and optical absorption, phase transition materials, and optical design using AI; the second covers research on thermophovoltaic elements, energy harvesting, and radiative cooling; and the third introduces research on photothermal materials' applications, such as solar steam generation, desalination, recyclable inks, and radiative textiles. Each chapter is authored by an expert whose research is focused on a specific related technology or application. Readers can apply the information in this book to address many common problems related to environment and energy conservation.

This book is invaluable for researchers and graduate students working in the fields of nanophotonics, energy, and environmentally friendly solutions, whether they are working on advancing the underlying technologies or expanding the range of usable applications to solve common global problems related to energy use, cooling, and resource consumption.

Kotaro Kajikawa is a Professor in the Department of Electrical and Electronic Engineering at the Tokyo Institute of Technology. He obtained his Bachelor's, Master's, and Doctorate degrees from Tokyo Tech. Professor Kajikawa is a member of The Japanese Society of Applied Physics and serves on the editorial board of Scientific Reports.

Junichi Takahara is a Professor of the Graduate School of Engineering at Osaka University, Japan. He gained his Bachelor's, Master's, and PhD all from Osaka University.

Thermal Plasmonics and Metamaterials for a Low-Carbon Society

Edited by
Kotaro Kajikawa and
Junichi Takahara

CRC Press
Taylor & Francis Group
Boca Raton London New York

CRC Press is an imprint of the
Taylor & Francis Group, an **informa** business

Designed cover image: The earth at night was held in the hands of humans. Earth day. Energy saving concept. Elements of this image furnished by NASA

MATLAB® and Simulink® are trademarks of The MathWorks, Inc. and are used with permission. The MathWorks does not warrant the accuracy of the text or exercises in this book. This book's use or discussion of MATLAB® or Simulink® software or related products does not constitute endorsement or sponsorship by The MathWorks of a particular pedagogical approach or particular use of the MATLAB® and Simulink® software.

First edition published 2024
by CRC Press
2385 NW Executive Center Drive, Suite 320, Boca Raton FL 33431

and by CRC Press
4 Park Square, Milton Park, Abingdon, Oxon, OX14 4RN

CRC Press is an imprint of Taylor & Francis Group, LLC

© 2024 selection and editorial matter, Kotaro Kajikawa and Junichi Takahara; individual chapters, the contributors

ISBN: 978-1-032-52904-2 (hbk)
ISBN: 978-1-032-52908-0 (pbk)
ISBN: 978-1-003-40909-0 (ebk)

DOI: 10.1201/9781003409090

Typeset in Minion
by SPi Technologies India Pvt Ltd (Straive)

Contents

Contributors

Muluneh G. Abebe is a FNRS Postdoctoral associate in the Micro- and Nanophotonic Materials Group at the University of Mons, Place du Parc, Belgium.

Ken Araki is a Postdoctoral Researcher at Arizona State University.

Takashi Asano is an Associate Professor in the Department of Electronic Science and Engineering, Kyoto University.

Rihab Benlyas is with the Department of Mechanical Engineering, Tohoku University.

Christopher Y. H. Chao is the Vice-President (Research and Innovation), Founding Director of the Policy Research Center for Innovation and Technology and Chair Professor of Thermal and Environmental Engineering at The Hong Kong Polytechnic University.

Alice De Corte is a PhD student in the Micro- and Nanophotonic Materials Group at the University of Mons, Place du Parc, Belgium.

Andrea Fratalocchi is a full Professor of Electrical and Computer Engineering and Affiliate Professor of Applied Mathematics and Computational Science at King Abdullah University of Science and Technology, Saudi Arabia.

Takuya Inoue is an Assistant Professor in the Department of Electronic Science and Engineering, Kyoto University.

Satoshi Ishii is Team Leader of the Optical Nanostructure Team at the National Institute for Materials Science in Japan.

Kotaro Kajikawa is a Professor in the Department of Electrical and Electronic Engineering at the Tokyo Institute of Technology.

Wakana Kubo is a Professor in the Division of Advanced Electrical and Electronics Engineering, Tokyo University of Agriculture and Technology.

Bjorn Maes is professor and group leader of the Micro- and Nanophotonic Materials Group at the University of Mons, Place du Parc, Belgium.

Susumu Noda is a full Professor with the Department of Electronic Science and Engineering and a director of Photonics and Electronics Science and Engineering Center (PESEC), Kyoto University.

Atsushi Sakurai is an Associate Professor with the Mechanical Systems Engineering Program, Department of Engineering, at Niigata University and a Research Professor with Niigata University Research Promotion Organization.

Makoto Shimizu is an Associate Professor with the Graduate School of Engineering at Tohoku University.

Junichi Takahara is a Professor of the Graduate School of Engineering at Osaka University, Japan.

Osamu Takayama is a Senior Researcher at the Department of Electrical and Photonics Engineering at the Technical University of Denmark.

Ross Y. M. Wong is a Postdoctoral Researcher at the Research Center for Materials Nanoarchitectonics of the National Institute for Materials Science in Japan.

Fei Xiang a PhD student in the Primalight Group in the Faculty of Electrical and Computer Engineering at King Abdullah University of Science and Technology (KAUST).

Hiroo Yugami is a Professor in the Department of Mechanical Engineering, Tohoku University.

Richard Z. Zhang is an Assistant Professor in the Department of Mechanical Engineering at the University of North Texas, Denton, TX.

CHAPTER 1

Toward Thermal Photonics by Integrating Thermal Engineering and Photonics

Atsushi Sakurai

Niigata University, Niigata, Japan

1.1 INTRODUCTION

What do you, the readers of this book, think of when you hear the word "thermal engineering" research, which is completely different from your field of study? I think you have a vague image of research on car engines or gas turbines. Certainly, these are important topics in mechanical engineering. It may seem that photonics and thermal engineering have little to do with each other, but this is not the case. For example, advanced laser measurement techniques are used to measure temperature and flow characteristics inside engines,[1,2] and laser spectroscopy techniques are also applied to various thermophysical property measurements used in mechanical design.[3,4]

"Thermal radiation" is included in the field of mechanical engineering, especially thermal engineering. It is one of the three forms of heat transfer, i.e., conduction, convection, and radiation, and is a physical phenomenon that must be considered in the design of high-temperature machines, such as engines, because its effect becomes more significant at higher temperatures. Recently, thermal photonics has become a new field of research within thermal engineering, especially at the boundary with photonics, and I have included the term "thermal photonics" in the title of this chapter. This term is taken from a review article by Professor Shanhui Fan of Stanford University.[5]

In the field of thermal radiation research, studies on the control of thermal radiation using artificial materials called metamaterials have been active for about 10 years.[6,7] Metamaterials were originally developed in the field of photonics[8] and the concept of metamaterials has been applied to this field. Now, the hot topics are non-equilibrium radiation[9] and the use of topological materials[10] to control energy transport and thermal radiation, which were unthinkable in the past. In this chapter, the authors present their research on thermal radiation control by metamaterials and their current work on thermal photonics power generation using non-equilibrium radiation.

1.2 WAVELENGTH-SELECTIVE CONTROL OF THERMAL RADIATION WITH METAMATERIALS

The word "metamaterial" is so common in the field of photonics that it does not need to be explained in detail. Metamaterials are artificially created materials with special physical properties that do not exist in natural materials. Metamaterials consist of an array of periodic meta-atoms with a fine nano- and microstructure, and this structure is designed to be able to manipulate electromagnetic waves.[8] However, it is still technically difficult to arrange meta-atoms in a three-dimensional array. On the other hand, it is relatively easy to integrate meta-atoms only on the surface, so the two-dimensional version of metamaterials is called a metasurface[11] (Figure 1.1).

Therefore, I have performed theoretical calculations on the control of thermal radiation using metasurfaces.[12] In the past, studies of thermal radiation in mechanical engineering have mainly focused on thermal energy transport, i.e., how photons originating from thermal radiation are emitted from a surface or space and transported to other locations, and the wave-like nature of thermal radiation has rarely been considered. In this case, the governing equations are the radiative transfer equations, which do not consider wave effects in detail because the system is much larger than the wavelength of the thermal radiation (although wave effects can be included in the scattering phase function). However, when the physical system under consideration is as large as the wavelength of the thermal radiation, as in the case of metasurfaces, wave effects must be taken into account as a matter of course, and the governing equations are Maxwell's equations. Thus, it was

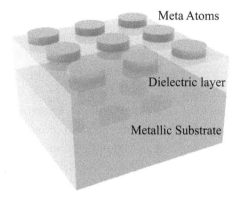

Meta Atoms

Dielectric layer

Metallic Substrate

FIGURE 1.1 Concept of metasurface. Metamaterials consist of an array of periodic meta-atoms with a fine nano- and microstructure. It is relatively easy to integrate meta-atoms only on the surface, so the two-dimensional version of metamaterials is called a metasurface.

necessary to combine thermal engineering and photonics to deal with thermal radiation phenomena at the nano- and micro-scale.

With the cooperation of experimental researchers who were interested in our theoretical calculations, we were able to obtain research results with good agreement between Finite Difference Time Domain (FDTD) calculations and experimental results.[13] Figure 1.2 shows a large-area metasurface and its spectral measurement results. Here we explain why it is possible to selectively extract thermal radiation only at specific wavelengths by using a metasurface. The metasurface contains a metallic part of aluminum and a dielectric part of CeO_2. If only aluminum is used, the

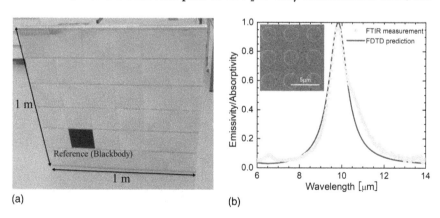

(a)

(b)

FIGURE 1.2 (a) Photograph of $Al/CeO_2/Al$ large-area metasurface, (b) Spectral measurements by FTIR and calculations of emissivity/absorptivity by FDTD calculation (inset: SEM image of meta-atoms).

emissivity should be zero according to Kirchhoff's law, since the reflectance of infrared light is close to 100%. However, experimental results show that the emissivity is close to 1. This is because the metal-insulator-metal (MIM) structure is a meta-atom that resonates with light of a certain frequency. In this case, the meta-atom is known to act as a kind of optical antenna because the electric current in the metal is amplified along with the amplification of the magnetic field based on Ampere's law. This phenomenon is often described as Magnetic Polariton.[14] In collaboration with JAXA's Institute of Space and Astronautical Science (ISAS), they aim to use the MIM metasurface as a radiator for infrared astronomical satellites and next-generation asteroid probes.[15]

1.3 OPTIMAL DESIGN OF ULTRA-NARROW-BAND THERMAL RADIATION EMITTERS USING MACHINE LEARNING

In recent years, machine learning has become an ubiquitous tool, but when Dr. Demis Hassabis of Google released AlphaGo a few years ago, machine learning research in materials design was still in its infancy. Even in metasurface research, the number of candidate material combinations and structural parameters was nearly infinite, making optimal design a challenge. An idea to solve this problem by applying machine learning to the design of thermal radiation emitters was proposed.[16] Figure 1.3 shows the design flow using machine learning.

The goal of this study was to design the optimal combination of materials and film thickness in a multilayer structure. At first glance, this study may seem simple because other optimal design methods, such as genetic algorithms, have been used in the past, but it turns out that this is not the case. For example, if three types of materials are arranged in N layers, the number of candidates is simply 3^N. Five or so layers are not a problem, but when 10 or 20 layers are used, the number of candidate combinations becomes so large that it is called combinatorial explosion. At this scale, a method such as machine learning, in which the algorithm itself learns the optimal path to reach the goal in the shortest time, becomes effective.

In this study, we designed how to arrange three materials, SiO_2, Al_2O_3, and Si, to produce narrow-band thermal radiation in the wavelength bands of 5, 6, and 7 μm. In this case, there are about 8 billion candidates; but by using machine learning, we were able to perform the optimal calculation with a computational cost of about 0.5% of the total number of candidates. The result is shown in Figure 1.4(a). Today, machine learning

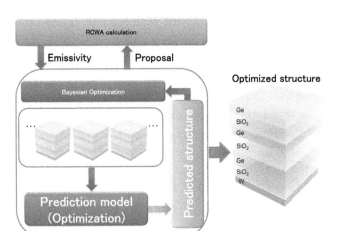

FIGURE 1.3 Design flow of ultra-narrowband thermal radiation emitter using machine learning. Based on the emissivity results calculated by Rigorous Coupled Wave Analysis (RCWA) method, candidate structures are estimated by Bayesian optimization, and machine learning is repeated to search for the optimized structure.

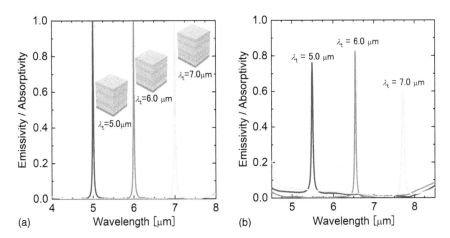

FIGURE 1.4 (a) Optimal structure and predicted emissivity spectrum by machine learning, (b) spectral measurements of multi-layered structures. Although the horizontal axis is different, both have very sharp emissivity peaks. The reason for the wavelength band shift is that the dielectric function used in the calculations are different from those of the fabricated films.

algorithms have advanced greatly, and it is now possible to compute the optimal number of candidates in the trillions, if not billions, of units.

What was new in this study was not only the optimal design based on machine learning, but also the actual fabrication of thermal radiation emitters based on that design. Since the multilayer structure can be fabricated by the sputtering method, we were able to observe a narrow-band thermal radiation spectrum, as shown in Figure 1.4(b). Although the wavelength band is gradually shifted to the longer wavelength side due to the difference in the calculated dielectric function, we succeeded in developing a narrow-band thermal radiator with an experimental Q-factor approaching 200 and were able to experimentally demonstrate the effectiveness of machine learning.

1.4 POSSIBILITY OF THERMAL PHOTONICS POWER GENERATION

Next, we challenged the academic question, "Can the blackbody radiation limit be exceeded even in the far field?" It was known that the blackbody radiation limit can be exceeded in the near field by Evanescent waves and phonon polariton resonance,[17] but whether it can be exceeded in the far field was unknown.[18]

In understanding thermal radiation, we have considered treating it as a "wave" rather than a "photon", but now we are back to "photons". The origin of thermal radiation is explained as follows. Carbon dioxide, for example, is composed of one carbon atom and two oxygen atoms, and its vibrational mode causes charge polarization in the molecule. When the molecule vibrates, the polarized charge vibrates with it, as if it were a source of electric current.[19] In other words, carbon dioxide molecules, like MIM metasurfaces, have optical antennas that are the source of thermal radiation. This is a correct electromagnetically based explanation of thermal radiation, but there is a physical phenomenon that cannot be understood with this explanation. This is non-equilibrium radiation in semiconductor materials.

In semiconductors, there is an energy state where electrons cannot exist, or a band gap, which causes the absorption and emission of photons. From the perspective of semiconductor engineering, thermal radiation in a thermal equilibrium system can now be viewed from a different angle. In any material, electrons can be in a variety of electronic states: they can receive thermal energy and jump like a whack-a-mole to the upper energy state, or they can emit photons of thermal radiation and fall to the lower energy state. Thermal equilibrium systems are based on this sensitive balance.

In semiconductors, as mentioned above, there is a band gap, and electrons that fall into the band edge are emitted as photons due to recombination between electrons and holes. Before falling into the band gap, electrons are thermalized by collisions with the surrounding lattice or by losing energy as thermal radiation photons. The probability of such photon emission increases dramatically when a non-equilibrium state is created by current injection, and this series of photon emission processes can be called non-equilibrium radiation.

In order to find out how this can be applied to thermal engineering, I would like to describe the transition of my previous research activities. The first goal was to investigate whether super-Planckian thermal radiation transport beyond the blackbody limit is possible in the far field. The blackbody limit is a theoretical limit that cannot be exceeded because it is the maximum value of blackbody radiative energy that can be derived from Planck's law for all matter. Unfortunately, studies that began with reckless attempts to exceed this limit have ended in disappointment.

For example, as shown in Figure 1.5, if nano- and micro-scale structures can be successfully brought into an optical resonance state, the light

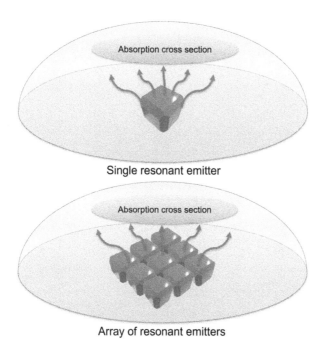

FIGURE 1.5   Absorption cross section for emitters in a single resonant state and for an array of multiple resonant emitters. In a single resonant emitter, the absorption cross section is larger than the geometric cross section, but in a resonant emitter array, they cannot be superimposed, so the array area and geometric cross section become closer.

absorption cross section in each structure will be larger than the geometric cross section, and the energy transport is expected to increase by "Planck radiation × absorption cross section". Note, however, that this does not mean that the Planck radiation is exceeded, only that the apparent cross section is increased. Then, even if one tries to extend the array of these nano- and micro-scale structures to the macroscopic scale, a superposition of individual resonance states is not possible, which is not possible in thermal equilibrium. In other words, the absorption cross section and the geometric cross section are close together, indicating that super-Planckian radiative heat transfer is not possible. A more detailed explanation can be found in the reference paper.[20]

Therefore, we changed our research strategy and focused on the non-equilibrium radiation described above. It is clear that non-equilibrium radiation (not super-Planck radiation) is generated by current injection, and thermal photonics (TPX) has shown one possibility to address the question of how to apply it to thermal engineering.[9] As shown in Figure 1.6, thermal photonics power generation has the same system configuration as conventional thermophotovoltaic (TPV) power generation, except that a non-equilibrium radiation emitter is used in the emitter part.[21,22] The current

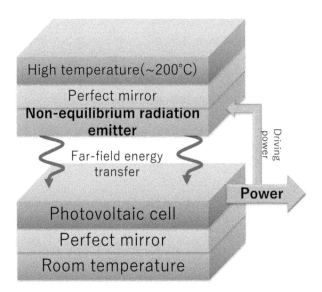

FIGURE 1.6 Schematic diagram of TPX power generation. A heat engine that operates by using a portion of the power generated by a photovoltaic cell to drive a non-equilibrium radiation emitter.

source for driving the non-equilibrium radiator is provided by a portion of the electricity generated by the photovoltaic cells, just as the compressor in a gas turbine engine is driven by a portion of the turbine output.

In this study, we focus on black phosphorus, a two-dimensional material with a band gap in the mid-infrared region, as a non-equilibrium radiation source and photovoltaic cell.[23] In the mid-infrared, compound semiconductors such as AlGaAs and InSb also have a band gap, but they are rare materials and thus expensive to produce. On the other hand, two-dimensional materials can be stacked by van der Waals forces and are attracting attention as low-cost, next-generation semiconductors.

The non-equilibrium radiation energy spectrum follows the generalized Planck law expressed by Equation 1.1.[24]

$$E_\lambda = \frac{\hbar\omega^3}{2\pi c_0} \frac{1}{4\pi^2} \int_0^\infty \xi(\omega,\beta)\beta \frac{1}{\exp\left(\dfrac{\hbar\omega-\mu}{k_B T}\right)-1} d\beta \tag{1.1}$$

where $\hbar$ is the reduced Planck constant, c_0 is light speed, ξ is transmission coefficient, β is wave vector, ω is angular frequency, μ is chemical potential, k_B is Boltzmann constant, T is temperature.

The generalized Planck's law allows non-equilibrium radiation and thermal radiation to be expressed in a unified equation by expressing the probability distribution of photons as a Bose-Einstein distribution. To solve this equation, we used a fluctuational electromagnetic simulation. This numerical method can directly represent the absorption and emission of photons inside a material.[25]

Figure 1.7 shows the results of the analysis of the light energy transport between two objects using this method. The temperature conditions were 450 K for the high temperature side and 300 K for the low temperature side. When an electric current is injected into the non-equilibrium radiation emitter using black phosphorus, the chemical potential inside the material changes, allowing energy transport in the optical frequency band above the band gap, which is larger than that of blackbody radiation.

Next, the performance as TPX power generation is evaluated. For comparison, the case where no current is injected into the non-equilibrium radiation emitter side, i.e., conventional TPV power generation, is used as a reference. The results are shown in Figure 1.8. The results show that the

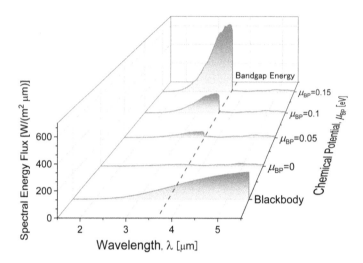

FIGURE 1.7 Effects of chemical potential on energy flux in TPX systems. In the wavelength range above the semiconductor band gap, energy transport beyond the blackbody limit is possible.

TPX power generation has about 10 times higher power density under the same temperature conditions. The thermal efficiencies normalized by Carnot efficiencies were about 20–25% for both types of power generation. It should be noted that these results are theoretical upper limits and that in reality the effect of non-radiative recombination is significant.

Currently, the waste heat energy in Japan is estimated to be 700 PJ per year, which is equivalent to the thermal energy emitted by 18 nuclear power plants. Approximately 70% of this waste heat is low-temperature waste heat (below 200°C), which is extremely difficult to recover. TPV power generation is suitable for power generation using thermal radiation energy from materials with a high temperature of nearly 2000°C. However, the application of such high-temperature heat sources is limited. Thermoelectric power generation can also use low-temperature waste heat, but it is difficult to achieve the efficiency and temperature difference with the heat source in the contact method. On the other hand, in TPX power generation, the emitter and receiver can in principle be separated by any distance. Of course, due to the view factor, it will be necessary to add a directional radiative transport in the future. However, it is easy to create a temperature difference in the non-contact method, and it will also be possible to achieve higher efficiency by concentrating the non-equilibrium radiation on the photovoltaic cell.

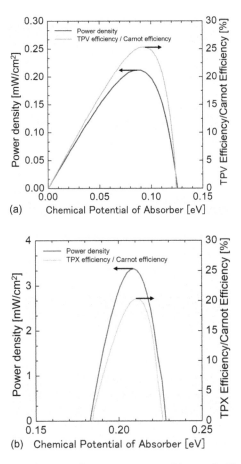

(a)

(b)

FIGURE 1.8 (a) Power density of TPV generation. Thermal efficiency normalized by Carnot efficiency; (b) Power density of TPX power generation under the same conditions. Thermal efficiency normalized by Carnot efficiency. The thermal efficiencies are not significantly different between both, but the power density of TPX power generation at the same temperature range is about 10 times greater than that of TPV power generation.

1.5 CONCLUSION

Traditionally, the study of thermal radiation in the field of thermal engineering has focused on the transport of energy between objects or within a space. However, there are now attempts to gain a deeper understanding of the origin of thermal radiation and to control the properties of thermal radiation itself, and the field has evolved as a boundary area combining thermal engineering and photonics. Originally, thermal radiation is naturally emitted light derived from thermal energy, which is always emitted from all materials beyond a temperature of 1 K. Therefore, thermal radiation is an abundance of energy around us. I would be very happy if readers of this chapter showed some interest in thermal radiation studies.

REFERENCES

1. K. Akihama, T. Asai, S. Yamazaki: *Applied Optics* **32** (1993) 7434–7441.
2. M. Matsuda, T. Yokomori, M. Shimura, Y. Minamoto, M. Tanahashi, N. Iida: *International Journal of Engine Research* **22** (2021) 1512–1524.
3. J.F. Torres, A. Komiya, E. Shoji, J. Okajima, S. Maruyama: *Optics and Lasers in Engineering* **50** (2012) 1287–1296.
4. Y. Nagasaka, T. Hatakeyama, M. Okuda, A. Nagashima: *Review of Scientific Instruments* **59** (1988) 1156–1168.
5. S. Fan: *Joule* **1** (2017) 264–273.
6. N.I. Landy, S. Sajuyigbe, J.J. Mock, D.R. Smith, W.J. Padilla: *Physical Review Letters* **100** (2008) 207402.
7. H. Wang, L. Wang: *Optics Express* **21** (2013) A1078–A1093.
8. J.B. Pendry, D. Schurig, D.R. Smith: *Science* **312** (2006) 1780–1782.
9. T. Sadi, I. Radevici, B. Behaghel, J. Oksanen: *Solar Energy Materials and Solar Cells* **239** (2022) 111635.
10. C. Guo, V.S. Asadchy, B. Zhao, S. Fan: *eLight* **3** (2023) 2.
11. N. Yu, F. Capasso: *Nature Materials* **13** (2014) 139–150.
12. A. Sakurai, B. Zhao, Z.M. Zhang: *Journal of Quantitative Spectroscopy and Radiative Transfer* **149** (2014) 33–40.
13. Y. Matsuno, A. Sakurai: *Optical Materials Express* **7** (2017) 618–626.
14. B.J. Lee, L.P. Wang, Z.M. Zhang: *Opt. Express* **16** (2008) 11328–11336.
15. S. Tachikawa, H. Nagano, A. Ohnishi, Y. Nagasaka: *International Journal of Thermophysics* **43** (2022) 91.
16. A. Sakurai, K. Yada, T. Simomura, S. Ju, M. Kashiwagi, H. Okada, T. Nagao, K. Tsuda, J. Shiomi: *ACS Central Science* **5** (2019) 319–326.
17. T. Inoue, K. Ikeda, B. Song, T. Suzuki, K. Ishino, T. Asano, S. Noda: *ACS Photonics* **8** (2021) 2466–2472.
18. D. Thompson, L. Zhu, R. Mittapally, S. Sadat, Z. Xing, P. McArdle, M.M. Qazilbash, P. Reddy, E. Meyhofer: *Nature* **561** (2018) 216–221.
19. Z.M. Zhang, *Nano/Microscale Heat Transfer*, Springer Cham, 2020.
20. Y. Xiao, M. Sheldon, M.A. Kats: *Nature Photonics* **16** (2022) 397–401.
21. A. Sakurai, Y. Matsuno: *Micromachines (Basel)* **10** (2019).
22. A. LaPotin, K.L. Schulte, M.A. Steiner, K. Buznitsky, C.C. Kelsall, D.J. Friedman, E.J. Tervo, R.M. France, M.R. Young, A. Rohskopf, S. Verma, E.N. Wang, A. Henry: *Nature* **604** (2022) 287–291.
23. Y. Matsuno, N. Nagumo, M. Araki, K. Yada, K. Yamaga, A. Sakurai: *Journal of Quantitative Spectroscopy and Radiative Transfer* **288** (2022) 108271.
24. P. Würfel, S. Finkbeiner, E. Daub: *Applied Physics A* **60** (1995) 67–70.
25. K. Chen, B. Zhao, S. Fan: *Computer Physics Communications* **231** (2018) 163–172.

CHAPTER 2

Angular Selective Multiband Emission Based on Optical Resonances Coupled with a Leaky Mode

Makoto Shimizu, Rihab Benlyas, and Hiroo Yugami

Tohoku University, Sendai, Japan

2.1 INTRODUCTION

Thermal radiation is spontaneous emission and is generally treated as incoherent light. Since thermal radiation is incoherent, it has a broad spectrum and almost an isotropic emission property. In accordance with this understanding, people have thought of radiant heat transport in thermal and mechanical engineering up to now. On the other hand, it is known that the coherence of thermal radiation can be improved by coupling it with surface modes through microstructures, and that it is possible to provide characteristics completely different from those of general thermal radiation, which has a narrow bandwidth and directional emission characteristics. In other words, the coherence control of thermal radiation increases the degree of freedom in the control of radiative heat transport, which has been conventionally considered, and enables advanced thermal management that was not possible with the conventional way of thinking.

By increasing the emissivity in areas of interest and decreasing it in less relevant regions, thermal radiation can be used to the fullest. As such,

DOI: 10.1201/9781003409090-2

13

spectral shaping has attracted a lot of attention. By manipulating the material properties and through judicious couplings, spectral shaping can be achieved. Metals and semiconductors emerge as one of the most commonly used materials for thermal radiation control as they support various resonance modes, such as surface plasmon polariton as well as localized surface plasmon resonances that can be used in the shaping of thermal radiation.

Spectral shaping has been studied thoroughly [1–8], whereas angular shaping remains challenging. Nevertheless, there exist several ways for angular control of thermal radiation. It has been proven previously that a thin film of polymers can achieve strong absorption at high angles by adjusting the thickness [9]. The mechanism behind this angular selectivity is the coupling of the molecular vibration of the polymers with a leaky mode that appears for thin films at high angles. Although quite simple and easy to integrate, the dependence on the molecular vibration makes this mechanism narrowband. Moreover, polymers cannot reach high temperatures, which limits the applications. Yagi-Uda antennas [10], where arrays of gold nanorods were deposited on SiO_x film, can also achieve angular shaping. Constructive interference in one direction and destructive in the other is used so as to obtain a phase coherent emission. Another example would be a single graphene sheet [11] on top of a dielectric spacer and on top of gold, that successfully reached perfect absorption in the mid-infrared (IR) range in high angles. When the critical coupling condition is satisfied, the absorption reaches 1. This coupling is achieved at a high angle called critical angle. Epsilon-near-zero (ENZ) [12–14] are artificial materials that can sustain permittivity below that of the free space and near 0 values, and can be used to achieve strong absorption at specific angles. Berreman leaky modes [15–18] where bulk polaritons are excited in thin films produce absorption at high angles of which the wavelength range depends on the thickness. Both ENZ and Berreman leaky modes make use of materials properties, and their intrinsic excitations are also known to achieve strong absorption at high angles. Consequently, they have a limited bandwidth. The current state of angular shaping of thermal radiation is that it can only be used for limited bandwidth, which limits the emissive power intensity on a considerable wavelength range. The wavelength region also depends highly on the characteristics and properties of the materials. In applications where only narrowband and angular shaping is needed, the existent technologies are enough. However, the research hasn't progressed enough yet to cover the case where a strong, directive emissive power on a wide range is needed.

Recently, relatively broadband (8–14 μm) emission and angular shaping is achieved through gradient ENZ structures in the far-IR range [19]. However, the dependency on the intrinsic property of materials makes flexible multiband or broadband control of thermal radiation difficult. A possible solution is to design multiple resonances that would couple with these angular-dependent modes, which would produce a multiband and angular-selective emitter with a large emissivity at high angles. A multiband and angular selective emitter is expected to contribute to energy saving through efficient cooling and energy generation such as thermophotovoltaics. Although the radiative heat transfer contribution is small at room temperature range, it increases as the temperature increases, and the impact of directive radiative cooling would increase as the temperature increases too.

Ferrell-Berreman modes have been previously used to achieve strong absorption at high angles, and consequently, emission at high angles. The mechanism behind this phenomenon is the coupling between the intrinsic excitation supported by the material (molecular vibration/phonon mode) with leaky modes. As the intrinsic excitation depends strongly on the material, Ferrell-Berreman modes offer in general a limited bandwidth and the range of the excitation cannot be controlled. The idea would be to artificially design multiple excitation that would couple with the Berreman leaky modes [20].

2.2 DESIGN OF MULTIBAND ANGULAR SELECTIVE EMITTERS

This chapter provides an explanation of the design concept behind multiband angular selective emitters. Furthermore, we present simulated results of these emitters, which are based on a metal-dielectric metal structure. The impact of each parameter on the angular selective properties is discussed. It is crucial to have a high fill-factor of the top metal in order to achieve both high angular selectivity and an increased number of emission peaks. Additionally, it is revealed that the relative thickness of the dielectric layer compared to the wavelength of interest plays a vital role in obtaining a large angular selectivity. This is because the angular selectivity exponentially decreases with an increase in the thickness of the dielectric layer.

2.2.1 Berreman Leaky Mode

Ferrell-Berreman modes usually refer to the excitation of leaky bulk polaritons. They were first studied by Ferrell [21] in 1958 for plasmon polaritonic thin-films in the UV range. In 1963, they were studied by Berreman [22] for phonon polaritonic thin films in the mid IR range. It was later proved that both mechanisms are the same, leading to the appellation of Ferrell-Berreman modes. They lie within the light cone as can be described in Ref. [12]. They are different from surface plasmon polariton, where the excitation is on the top surface. For Ferrell-Berreman modes, the energy propagates within the bulk volume of the metal. They have also shown adaptable position within the dispersion curve, and are characterized by absorption from the visible to IR range, depending on the materials and layer thicknesses. Strongly affected by the surface, these modes have different applications from thin film characterization [23], imaging, sensing [24] and directional control of thermal radiation.

In 2012, Vassant [15] studied the Ferrell-Berreman modes on a structure made of a thin film of SiO_2 on gold. The Berreman-leaky mode appears for very thin films and couples with the phonon excitation supported by SiO_2 in the IR range. This particular mode is of interest for the design of the emitter. Here, a dielectric film of permittivity ε_2 and thickness d deposited on a metal of permittivity ε_3 illuminated in a medium of permittivity ε_1 is considered. The surrounding medium is considered air so that $\varepsilon_1 = 1$. The idea is to search for a TM mode of form $H_y \exp(iKx - ik_{z,1})$ in medium 1 and of the form $H_y \exp(iKx - ik_{z,1})$ in medium 3 where $k_{z,n} = (\varepsilon_n \omega^2/c^2)^{1/2}$ with the condition that $\text{Im}(k_{z,n}) + \text{Re}(k_{z,n}) > 0$. The dispersion relation of the TM mode is given by:

$$\left(1 + \frac{\varepsilon_1 k_{z,1}}{\varepsilon_3 k_{z,3}}\right) = i\tan\left(k_{z,2}d\right)\left(\frac{\varepsilon_2 k_{z,3}}{\varepsilon_3 k_{z,2}} + \frac{\varepsilon_1 k_{z,2}}{\varepsilon_2 k_{z,1}}\right) \tag{2.1}$$

Considering polaritonic materials, the analytical model of the permittivity is given by:

$$\varepsilon(\omega) = \varepsilon_\infty \frac{\omega^2 - \omega_L^2 + i\omega\Gamma}{\omega^2 - \omega_T^2 + i\omega\Gamma} \tag{2.2}$$

If we consider SiO_2, the phonon mode is supported frequency between 950 and 1250 cm^{-1} and $\varepsilon_\infty = 2.0955$, $\omega_L = 1220.8$ cm^{-1}, $\omega_T = 1048.7$ cm^{-1} and

$\Gamma = 71.4$ cm^{-1}. For a thickness of the dielectric film such that the two interfaces do not interact ($\text{Im}(k_z)d \gg 1$), we expect to find the dispersion relation of the surface phonon polariton at the air/dielectric interface and the dispersion relation of the surface plasmon at the metal/dielectric interface. This can be translated to the limit $K \to \infty$ so that $\tan(k_{z,2}d) \to i$, $k_z \to K$. The dispersion relation becomes:

$$\left(\varepsilon_1 + \varepsilon_2\right)\left(\frac{1}{\varepsilon_3} + \frac{1}{\varepsilon_2}\right) = 0 \qquad (2.3)$$

So, either the dispersion relation of the surface wave at the upper interface $\varepsilon_1 + \varepsilon_2 = 0$ or the dispersion relation at the lower interface $\varepsilon_2 + \varepsilon_3 = 0$ are obtained. The solutions correspond respectively to the surface phonon mode at the dielectric/air interface and to the surface plasmon at the metal/dielectric interface. These two equations give the asymptotes in the dispersion relation. Equation (2.1) is solved numerically for frequencies near ω_L. The quantity d' is introduced as $d' = \tan(k_{z,2}d)/k_{z,2}$ that tends to d when d tends to 0. Equation (2.1) then becomes:

$$\omega = \omega_{FB} + \frac{K^2\left(\omega_L{}^2 - \omega_T{}^2\right)}{\left(A^2 - K^2\right)\left(\omega + \omega_L{}^2/\omega_{FB}\right)} \qquad (2.4)$$

where $\omega_{FB} = \omega_L\sqrt{1 - \dfrac{\Gamma^2}{4\omega_L{}^2}} - i\dfrac{\Gamma}{2}$ is the complex solution of $\omega^2 - \omega_L{}^2 + i\omega\Gamma = 0$. With difference form, it can be written by:

$$A = \frac{i\varepsilon_\infty}{d'}\left(\frac{k_{z,1}}{\varepsilon_1} + \frac{k_{z,3}}{\varepsilon_3} - id'\left(\frac{\omega^2}{c^2} + \varepsilon_2\frac{k_{z,1}k_{z,3}}{\varepsilon_1\varepsilon_3}\right)\right) \qquad (2.5)$$

As shown in the dispersion relation with different thicknesses of SiO$_2$ plotted in Figure 2.1, the point A corresponds to the Berreman mode, B to the ENZ and the point C to the surface phonon mode.

As such, it can be assumed that the Surface plasmon polariton (SPP) couple with the far-field light through the coupling with the Berreman-leaky mode, which leads to the strong absorption at high angles. As mentioned above, they are however coupled with existing modes such as phonon or molecular vibrations. Such modes depend highly on the materials and

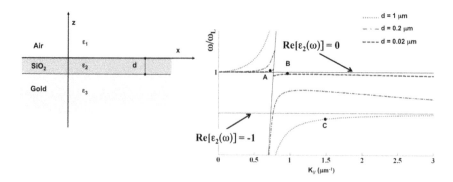

FIGURE 2.1   Berreman mode excited in thin SiO₂ film on gold from. The point A corresponds to the Berreman mode, B to the ENZ and the point C to the Surface phonon mode. (Reproduced with permission from Ref. [15].)

the considered structure, which limits the bandwidth, the choice of the range, and the intensity of the emissive power. Consequently, the applications are also quite limited. By the contrast, supposing a structure had multiple modes throughout the IR range that could couple with Berreman mode, the limitation on the bandwidth would be lifted. It would also result in multiband absorption at high angles, and consequently a strong emissive power at high angles.

2.2.2 Design of Emitters

Ferrell-Berreman modes have been previously used to achieve strong absorption at high angles, and consequently, emission at high angles. The mechanism behind this phenomenon is the coupling between the intrinsic excitation supported by the material (molecular vibration/phonon mode) with the Berreman-leaky modes. As the intrinsic excitation depends strongly on the material, this mode offers in general a limited bandwidth and the range of the excitation cannot be controlled. The idea would be to artificially design multiple excitations that would couple with the modes. SPPs are one of the candidates for designed resonance. When using a grating coupler, the SPP excitation can be designed by tuning the period and incident angle. As such, an SPP grating coupler is used on top of an SiO₂ thin film on top of back reflective gold, as can be seen in Figure 2.2(a). As such, the emitter consists of periodic gold gratings of 100 nm thickness in order to have complete reflection. They should not be too thick, where longitudinal modes would be dominant. The gratings are of width L with a period P deposited on a layer of SiO₂ of thickness d on top of a back reflective gold mirror.

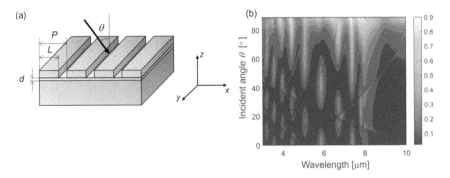

FIGURE 2.2 (a) Geometry of the emitter(absorber) considered in this study. A thin SiO$_2$ layer with thickness of d is coated on the gold mirror. On top of the SiO$_2$ layer, a one-dimensional gold grating with period of P and width of L is formed. In the simulation, the thickness of the grating is constant to be 0.1 µm. The incidence is from a direction perpendicular to the grating with a zenith angle of θ. (b) A contour map of the simulated absorptivity spectrum with different incident angle from 0° to 89° using the model with $P = 10$ µm, $L = 9$ µm, and $d = 0.1$ µm. (Reproduced with permission from Ref. [20].)

Maxwell's electromagnetic waves equations are solved in order to calculate the optical properties. However, in most cases, Maxwell equations cannot be solved analytically. As such, numerical methods are used. In this case, the Rigorous Coupled-Wave Analysis [25], (RCWA) method is chosen. It is an especially powerful method for modelling gratings. It is a semi-analytical technique, and picks one direction, for example (longitudinal), to solve analytically, and the other two (transverse) are solved numerically. RCWA uses a discrete Fourier transform to discretize the fields in the transverse direction. It represents the fields and materials as a set of plane waves. Both normal and total reflection, as well as total absorption, are simulated in each case. Simulated angular absorptance for an emitter of $P = 10$ µm, $L = 9$ µm and $d = 0.1$ µm is shown in Figure 2.2(b). As expected, the angular absorption shows high angular selectivity and is large at only high grazing angles, at almost 80° and above. The absorption is multiband, as it has numerous peaks throughout the range of study. The mechanisms behind these resonance peaks will be explained in detail in the following section.

2.2.3 Analysis of Resonance Mode

For explanation's sake, an emitter of the following dimensions $P = 5$ µm, $L = 4$ µm and $d = 0.1$ µm is chosen to study the resonance mechanisms,

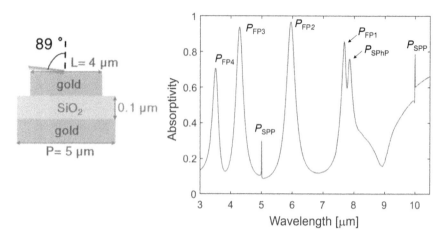

FIGURE 2.3 Simulation result of absorptivity using the model with $P = 5$ μm, $L = 4$ μm, $d = 0.1$ μm, and $q = 89°$ under TM polarization. Each peak is labelled according to the origin of the peak. P_{SPP} is the peaks derived from surface plasmon polariton and P_{SPhP} is the peak derived from surface phonon polariton. P_{FP} shows the peak derived from the coupling of the SPP to the FP-like resonance and their subscripts numbers are mode numbers. (Reproduced with permission from Ref. [20].)

as it shows a smaller number of absorption peaks than larger structures. As indicated in Figure 2.3, the absorption peaks are due to three different kinds of mechanisms. The first one is SPPs. The second one is due to the Berreman-leaky mode that couples with SPPs coupled with Fabry-Pérot resonances, and the third one is Berreman-leaky mode coupled with the phonon mode.

Bandwidth wise, we can globally distinguish between two kinds of peaks: very narrowband and broader peaks. The broader peaks are attributed to the SPPs/Fabry-Pérot resonances that will be discussed later. The narrowband peaks are associated with the SPPs resonance, and result of the excitation of SPPs by incident light on the top gold gratings. The excitation condition is given by the following equation.

$$P\left(\mathrm{Re}\left(k_{SPP}\right) + \frac{2\pi}{\lambda}\sin\theta \right) = 2\pi m \tag{2.6}$$

with

$$k_{SPP} = \frac{2\pi}{\lambda}\sqrt{\frac{\varepsilon_{air}\varepsilon_{air}}{\varepsilon_{air} + \varepsilon_{gold}}} \tag{2.7}$$

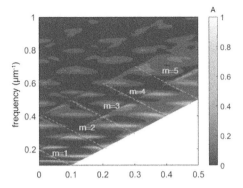

FIGURE 2.4   Dispersion relation for an emitter of P = 5 µm, L = 4 µm, d = 0.1 µm, and the corresponding SPPs wavelengths and their diffraction orders excited at 89.

that is given by the dispersion relation of the plasmons excited on the air/gold, with P being the period of the gratings and m the diffraction order. This equation is numerically solved and the resonant wavelengths and their corresponding diffraction orders are shown in the dispersion relation of the emitter (See Figure 2.4).

Fabry-Pérot resonances [26, 27] are at the core of numerous spectroscopic applications. In 1960, they have proven their ability to produce lasers [28], and their performance is increasing by the year. Typically consisted of two reflective mirrors separated by a vacuum or a chosen dielectric, the light interference results in a strong confinement, which make Fabry-Pérot devices good candidates as an absorber/emitter.

We assume that the broader peaks are due to Berreman mode that is excited by the SPPs and Fabry-Pérot coupling [29, 30]. To confirm that, the metal-insulator-metal (MIM) dispersion relation and the Fabry-Pérot resonance condition are solved. The coupling condition is given by:

$$\begin{cases} 2L\beta + \varphi_r = 2m\pi \\ \varepsilon_{SiO_2} k_{gold} + \varepsilon_{gold} k_{SiO_2} \tanh\left(k_{SiO_2} d/2\right) = 0 \\ \beta^2 - \varepsilon_{SiO_2} k_0^2 = k_{SiO_2}^2 \\ \beta^2 - \varepsilon_{gold} k_0^2 = k_{gold}^2 \end{cases} \quad (2.8)$$

Where φ_r is the phase shift of reflection and L is the width of grating and also the length of Fabry-Pérot resonance, k_0 is the wavevector of light in vacuum, and β is the complex propagation constant of the SPPs in an MIM waveguide. The first equation corresponds to the Fabry-Pérot resonance condition. The second equation correspond to the dispersion relation in an MIM structure. The third and fourth equations are substituted in the dispersion relation so that the propagation constant is obtained. Finally, the Fabry-Pérot matching equation is solved. The excited SPPs on the gold/SiO$_2$ interface couple with Fabry-Pérot resonances resulting in localized waves in the SiO$_2$ layer. Equation (2.8) is numerically solved, confirming the theory that the broader peaks at 3.51 μm, 4.30 μm, 5.95 μm, and 7.68 μm, are due to the SPPs/Fabry-Pérot coupling with diffraction orders of 4, 3, 2, and 1, respectively. Figure 2.5 shows the magnetic field (H_y) for incident electromagnetic waves at the resonant wavelengths at incident angle of 89°. The wave is localized and amplified in the SiO$_2$ layer and the number of nodes corresponds to the diffraction orders, a feature of Fabry-Pérot resonances. These peaks depend strongly on the thickness of SiO$_2$ as can be seen in Figure 2.6 where the absorption spectrum has been plotted for three thicknesses, 0.02 μm, 0.1 μm. It can be seen that for a very thin SiO$_2$ film $d = 0.02$ μm, the peaks are weaker and

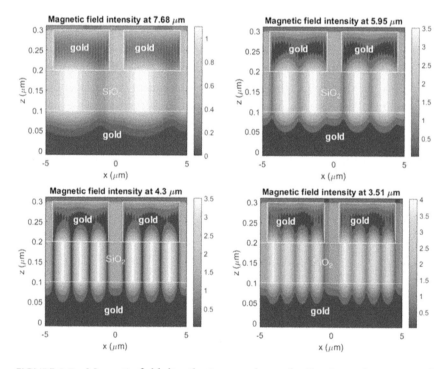

FIGURE 2.5 Magnetic field distribution are shown for $P = 5$ μm, $L = 4$ μm and SiO$_2$ thickness of $d = 0.1$ μm at resonant wavelengths with incident wavelengths of 7.68 μm, 5.95 μm, 4.30 μm, and 3.51 μm.

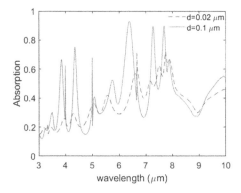

FIGURE 2.6 Comparison of absorption spectrums at 89° for two thicknesses d = 0.02 μm and d = 0.1 μm for an emitter of P = 5 μm, L = 4 μm.

broader as compared to d = 0.1 μm. The peak positions and shape also seem to be highly dependent on the thickness.

It is well known that SiO_2 has phonon mode [31, 32] at around 1100 cm^{-1} that is due to the Si-O-Si asymmetrical stretching bonds. For a very thin film of SiO_2 on gold substrate, the phonon mode couples with Ferrell-Berreman modes resulting in strong absorption at 1100 cm^{-1} range at high angles.

2.2.4 Influence of Geometrical Parameters on Angular Selectivity

In the previous section, the different mechanisms behind the absorption peaks are clear, the next step is to study the influence of the geometrical parameters of the emitter, i.e., the width of the gratings and the thickness of SiO_2.

The angular absorption is computed for an emitter of a period P = 15 μm, SiO_2 thickness d = 0.1 μm, and the width L is varied from 10 μm to 14 μm, as shown in Figure 2.7. As the width increases, so does the angular selectivity and the number of peaks. As the width increases, so does the fill factor that is defined as the fraction of the grating period that is filled with the grating material. This leads us to consider a mechanism of fill factor rather than the influence of the size of the period and the width separately, in the following section.

The influence of SiO_2 thickness d is studied next. The effect of varying the thickness of SiO_2 of an emitter of a period P = 10 μm and of width L = 9 μm, where the enhancement of absorption is between 80° and 10°, is plotted in Figure 2.8. The enhancement of absorption is defined as the ratio of average absorption between 80° and 10°, in the range of study 3

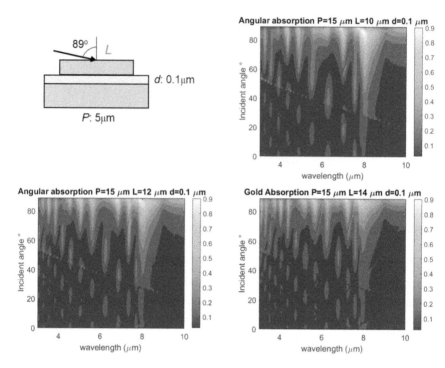

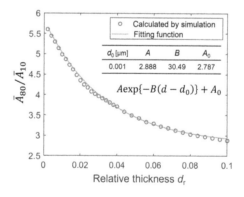

FIGURE 2.7 Angular absorption is shown for various emitters of same period 15 μm, same SiO_2 thickness 0.1 μm and different widths L of 10, 12, and 14 μm.

FIGURE 2.8 The ratio of average absorptivity, from 3 to 10 μm, between incident angles of 80° and 10° is shown as a function of relative thickness defined as $d_r = d/(5 \text{ μm})$. (Reproduced with permission from Ref. [20].)

μm to 10 μm. As explained before, the expected mechanism behind the strong absorption at high angles is the Berreman-leaky modes that appears for very thin films. As the thickness increases, the angular selectivity is weakened, and the absorption peaks become more angular independent.

This can be explained by the interaction of the SPPs of the top gold grating and the bottom gold mirror. These SPPs interact through the SiO_2 layer, and their interaction weakens as the thickness d increases. As the SiO_2 layer becomes thicker, the distance between the gold mirrors also increases. The Fabry-Pérot resonance is hence weaker. However, as it stands, similar structures [33–35] have used thinner SiO_2 films for angular independent applications. This brings us to consider a mechanism of relative thickness rather than the absolute thickness d of SiO_2. In our case, for the range of 3 μm to 10 μm we consider the wavelength of interest to be of 5 μm. The relative thickness is then defined by: Relative thickness = absolute thickness/wavelength of interest. By using the relative thickness, we could see the same exponential decay as in Figure 2.8, even with the different structures which have different structural parameters. Through this result, it is possible to compare with structures that have used thinner films, for applications in the visible and near IR range. For example, for a thickness of SiO_2 of $d = 0.04$ μm and a wavelength of interest standing at $d = 0.6$ μm, the relative thickness is about 0.066, where the emitter shows small angular selectivity, where the difference in the spectrum at high angle and low angle is already small at a relative thickness of 0.04.

The angular selectivity is supposed to appear for very thin films of SiO_2, consequently small relative thicknesses. However, it remains unclear up to this point whether there is an upper limit, and whether the fill factor influences the angular selectivity. As such, the enhancement of absorption between 80° and 10° as a function of fill factor and the relative thickness is shown in Figure 2.9. As it is a ratio of absorption, the volume effect is

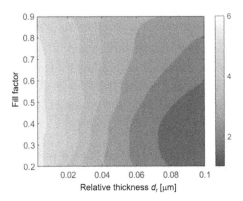

FIGURE 2.9 Factor of absorption enhancement between 80° and 10° incident light as a function of Fill factor and Relative thickness. (Reproduced with permission from Ref. [20].)

reduced, and the angular selectivity can be evaluated. It can globally be seen that for very thin SiO$_2$ films, consequently very small values of relative thickness, the absorption at 80° is up to 6 times higher than that at 10°, which proves the angular selectivity for thin films. The fill factor also seems to affect this angular selectivity. If we take a fill factor of 0.2, then the enhancement is up to 4 only up to a relative thickness of 0.03. However, for a fill factor of 0.9, it is up to 4 for up to a thickness of 0.07. The fill factor affects the number of peaks. Indeed, as the mirror gets larger, more Fabry-Pérot resonances can occur. These peaks couple with the Berreman-leaky modes at high angles, leading to a larger intensity as opposed to smaller mirrors, smaller fill factors.

To confirm that, the angular absorption for emitters of different fill factors for different periods is studied. It can be seen that the behavior is globally the same in Figure 2.10 whether it is for a period of 10 μm, 15 μm or 20 μm. As the fill factor increases, the number of peaks increases, as does the angular selectivity. It can be explained by the strong diffraction that occurs for large width, small slits. The number of peaks increases because of the Fabry-Pérot resonance mechanism, where the top gold grating and back reflective gold act like mirrors. As the width of the mirror increases, more Fabry-Pérot phenomenon can be observed.

Angular selectivity-wise, it can be concluded that it is the strongest for large fill factors and thin SiO$_2$ films, consequently small relative thicknesses. However, as we are talking about a ratio of average absorption, the intensity is not taken into account. As such, the influence of both fill factors and SiO$_2$ thickness on the intensity, consequently the intensity of the emissive power, is studied in the following section.

2.2.5 Influence of Geometrical Parameters on Emission Intensity

To evaluate the influence of SiO$_2$ thickness and fill factor on the emissive power, we consider different cases with a low fill factor $F = 0.1$ and large fill factor $F = 0.9$ and consider different thicknesses d, very thin at 0.03 μm, thin at 0.1 μm and thick at 0.5 μm. The emission intensity is calculated from Equation (2.9),

$$I(T,\theta)d\Omega = \int_\lambda \Theta(\lambda,T)\varepsilon(\lambda,\theta)d\lambda \qquad (2.9)$$

where Θ is the Planck function, ε the emittance that is equal to the absorptance, $d\Omega$ the solid angle and T the temperature. The emission

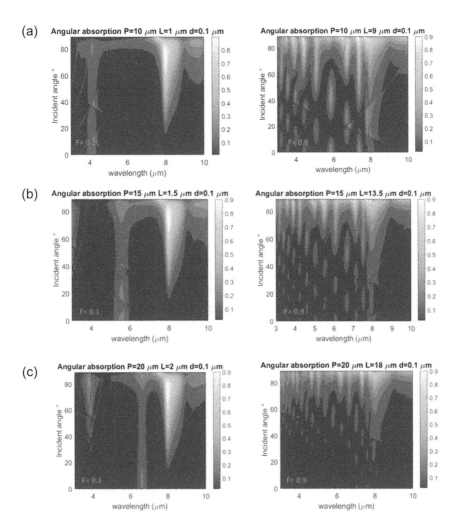

FIGURE 2.10   Angular absorption for various emitters of constant SiO$_2$ thickness $d = 0.1$ μm of relative thickness 0.02, different periods and fill factors. On the right side shows the result of $F = 0.9$, on the left side shows $F = 0.1$. (a) $P = 10$ μm, (b) $P = 15$ μm, (c) $P = 20$ μm. (Reproduced with permission from Ref. [20].)

intensity is then plotted between 3 and 10 μm for a temperature of 373 K as can be seen in Figure 2.11 where the angles are shown. As the emitter is polarization dependent, the result is multiplied by a factor of 0.5. For a fill factor of 0.1 the dominant resonant mode is the phonon mode that is the strongest for thin films. However, as the fill factor increases to 0.9, the dominant peaks are the modes derived from SPPs coupled with

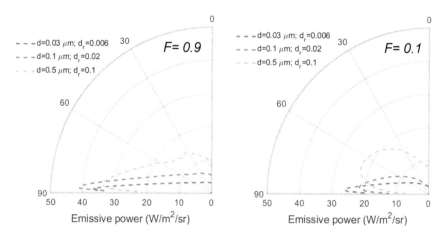

FIGURE 2.11 Simulation of the angular dependence of the average absorptivity from 3 to 10 μm using models with $P = 10$ μm, and $L = 1$ μm (left) or $L = 9$ μm (right). Three different SiO$_2$ thicknesses with $d = 0.03$, 0.10 and 0.50 μm are considered. The relative thicknesses are shown for each. (Reproduced from Ref. [20].)

Fabry-Pérot. When the SiO$_2$ thickness is very small, the reflection by the gold mirror becomes more dominant and the emitter becomes closer to a pure gold grating structure. As such, the strongest emission intensity is reached for a fill factor of 0.9 and a thickness $d = 0.1$ μm at a value of 41 W m^{-2} sr^{-1}. This value stands at 37 W m^{-2} sr^{-1} for $d = 0.03$ μm. On a side note, in order to limit the dependence on azimuthal angle, 2D gratings can also be considered. Figure 2.12 shows the simulation results of 2D gratings, with a fill factor of 0.9 for a thickness of SiO$_2$ of 0.1 and 0.5 μm. The simulation is run under unpolarized light. It can be seen that the multiband effect is present. Moreover, the influence of the relative thickness is also present.

Similar to the 1D case, the 2D case is multiband and angular selective, more so for small relative thicknesses (See in Figure 2.12). Although the emissive power is slightly lower under unpolarized conditions, it remains considerably large (24 W m^{-2} sr^{-1}) and of more flexible use thanks to its independence on an azimuthal angle. For both 1D and 2D emitters, we can conclude on the optimal parameters being large fill factors and relatively thin SiO$_2$ thicknesses. The 1D emitter, being the core of this study, will be studied experimentally in the following chapter. Its optimal parameters will be set at a thickness d around 0.1 μm corresponding to a relative thickness of 0.02, and a large fill factor of at least 0.8.

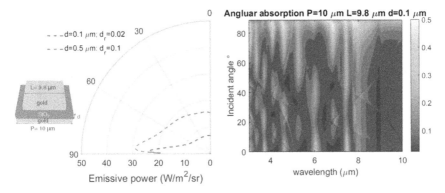

FIGURE 2.12 Simulation result of 2D structure cases which have square resonator on top of SiO₂ layer.

2.3 DEMONSTRATION OF THE MULTIBAND AND ANGULAR SELECTIVE EMITTER

In this chapter, we present an explanation of the experimental findings concerning the angular selective emitter we proposed. We assessed the angular absorption characteristics of fabricated emitters fabricated using photolithography to evaluate their emission properties in accordance with Kirchhoff's law. The measured results demonstrated a strong agreement with the simulated results, thus confirming the angular selective emission properties across multiple wavelength bands.

2.3.1 Emitter Fabrication

For micro and nano-patterning, etching and lift-off processes are the most commonly used. The fabrication steps can be seen in Figure 2.13.

The base substrate is chosen as a 2-inch Si wafer. First, the back reflective gold mirror is sputtered (Shibaura, CFS-4EP-LL). The thickness of gold is chosen as a minimum of 120 nm, so as to be sure to eliminate the transmission. In order to improve the adhesion of gold on Si substrate, 5 nm of chromium are sputtered. Then, about 5 nm of chromium are sputtered again, followed by the layer of SiO₂ of variable thickness. The sample is then prepared for laser lithography. First, the adhesion promoter HMDS is spin coated for 20 s at 3000 rpm, followed by a soft baking of 5 min at 115°C. Negative resist ZPN 1150 by Zeon is then spin coated for 20 s at 3000 rpm. The sample is then baked at 90°C for 90 s. Maskless Aligner MLA150 from Heidelberg Instruments is then used for laser-direct lithography. The sample is then developed using Tetramethylammonium

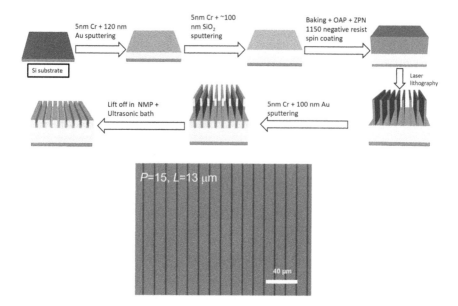

FIGURE 2.13 Fabricated process of the sample and an SEM image of the fabricated one with $P = 15$ μm and $L = 13$ μm.

Hydroxide (TMAH 2.38%). After that, about 5 nm of chromium and 100 nm of gold are sputtered. The thickness of the sputtered Cr is small enough to have negligible impact on the experiment, according to simulation results. Finally, the resist is lifted off using the resist stripper 1-Methyl-2-pyrrolidon (NMP). The fabricated grating structures can be seen in Figure 2.13.

2.3.2 Measurement of the Angular Selectivity

To measure the optical properties of the emitter, Fourier Transform Infrared Spectrometer (FTIR-6300, JASCO Japan) is used. It is purgeable as it includes a fully sealed and desiccated interferometer chamber and can support full or half vacuum with separate control of the interferometer, optical pathways, and sample compartment.

The FTIR equipment is used with attachments to measure normal reflectance at 10° (JASCO, RF-81S) and 80° (JASCO, RAS PRO410-B). A grid polarizer (JASCO PL-82, Wire grid polarizer KRS-5) is used, and the polarization angle can be changed from 0° to 180°, with 0° being TE polarization and 90° TM polarization. Also, to measure the normal reflectance from 55° to 85°, a variable angle attachment (JASCO PR-510i) is used. The sample's reflective energy is measured and the reflectance can be deduced.

Because of the presence of the back reflective mirror, the transmission is null. As such, we can obtain the absorption with $A = 1 - R$. As most of the used attachments measure normal reflection, a comparison needs to be made between both normal and total reflection, also referred to as specular and diffuse reflection. Indeed, as diffraction occurs, reflection and, consequently, absorption are strongly affected by it. Globally, for an emitter, that are made of large fill factors and thin SiO_2 films, consequently small relative thicknesses, the absorption spectrum from normal and total reflection were quite similar, with a slight difference in intensity. As such, for these dimensions, the diffraction is negligible and the measurement of normal reflectance can be used as an approximation for total reflectance. For all reflection measurements, the FTIR equipment is purged and brought to a vacuum state. Figure 2.14 shows the absorption spectrum of an emitter of $P = 15$ μm, $L = 13$ μm and $d = 0.15$ μm at high angle 80° and low angle 10°.

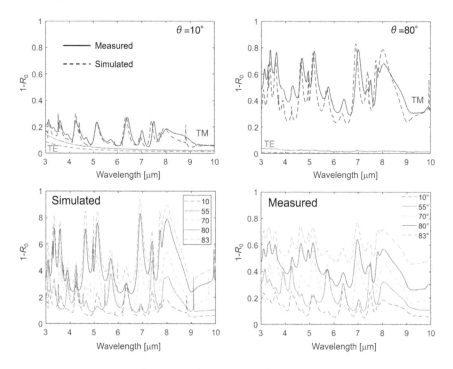

FIGURE 2.14 Measured spectra of 1-normal reflectance at the fabricated sample with different incident angles (10° and 80°) and different polarization. In the upper two graphs, dashed line shows simulated results and the solid line shows measured result. In the lower two graphs, simulated (left) and measured (right) intermediate incident angles are plotted.

It can be seen that the measured spectra are almost consistent with the simulated on, and the peak positions are almost the same. The slight difference can be attributed to the difference between the simulation model and the real sample, which includes the difference in optical properties of the materials and the presence of defects. What eventually stands out is that the sharp peaks previously attributed to the SPPs resonance are broader and weaker in intensity in reality. The SPPs depend strongly on the surface condition. Moreover, the simulation assumes an infinite perfect model, which can explain why the peaks in the simulation are so narrowband, whereas the resonance in the real emitter is weaker. However, the peaks attributed to the phonon mode, as well as the coupling between SPP and Fabry-Pérot, are present and agree quite well with the theory. Through this measurement, the dependence on the angle of incidence is clear. Indeed, the absorption intensity at low angle stands at an average of about 0.1 and increases to about 0.5 at the angle of 80°. The dependence on the polarization is also proven, where the emitter acts like a reflective mirror in TE polarization conditions with low absorption and high reflectance. The absorption spectrum at intermediate angles at 10°, 55°, 70°, 80°, 83° for emitter of $P = 15$ µm, $L = 13$ µm and $d = 0.15$ µm is measured, as can be also seen in Figure 2.14. Although broader because of defects and material differences, the peak positions remain coherent with the simulation. For angles up to 70° the absorption intensity remains quite small. It only gets significantly larger for angles >80°.

2.3.3 Dependence of Absorption Spectrum Intensity on SiO₂ Thickness

The influence of the thickness of SiO_2 on the intensity is then studied experimentally. For this purpose, the absorption spectrums of three emitters of the same size $P = 10$ µm, $L = 9$ µm and different thicknesses $d = 0.04$ µm, 0.11 µm and 0.5 µm are studied. When the thickness is increased from $d = 0.04$–0.11 µm, the absorption intensity also increases, as shown in Figure 2.15. This is coherent with the theory, where the strongest intensity/highest emissive power appears for moderately thin films of SiO_2 rather than very thin films, where the reflection becomes more dominant. However, what stands out is the strong absorption for a thick film of $d = 0.5$ µm, where the intensity is the strongest.

The absorption obtained from simulated normal reflection is also coherent with the experiment. This leads us to consider the diffraction

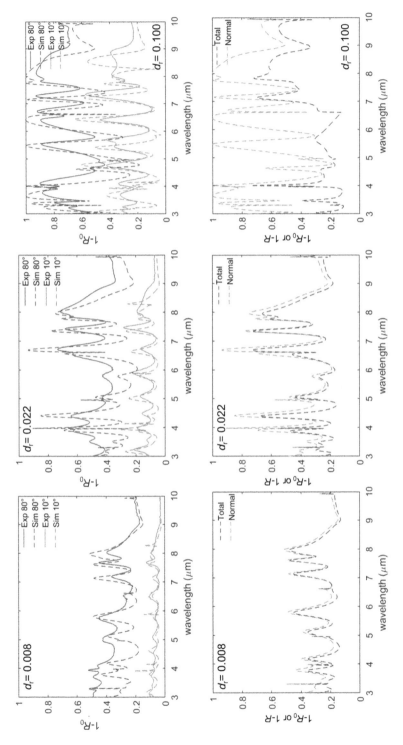

FIGURE 2.15 Measured spectra with different SiO$_2$ relative thicknesses (0.008, 0.022, 0.1) are shown in upper three graphs. The lower three graphs show the comparison between the simulated spectra of 1-normal reflectance (1-$R0$) and 1-total reflectance (1-R) with different SiO$_2$ relative thicknesses (0.008, 0.022, 0.1).

effect, that might not be negligible for thicker SiO_2 films. It can be seen that for both thicknesses of $d = 0.04$–0.11 μm the total and normal simulated absorption agree very well, with a slight difference in amplitude. However, for a larger thickness, the difference is no longer negligible. Nevertheless, the experimental normal reflectance is coherent with the simulated total reflection.

In order to evaluate the angular selectivity dependence on the relative thickness, the enhancement of absorption between an incident angle of $80°$ and $10°$ is calculated as can be seen in Figure 2.16. The measured angular selectivity agrees well with the simulated results, especially at small relative thicknesses. Thus, we can conclude experimentally on the effect of the relative thickness on the angular selectivity, that it is weaker for large thicknesses.

Finally, in order to clarify the difference from a conventional isotropic emitter, the radiative heat transport rate from the emission surface to the wall by varying the height of the wall are analyzed in the configuration shown in the inserted figure and results are summarized in Table 2.1. The heat transport efficiency is calculated by dividing the thermal radiation reaching the wall surface, which is calculated by ray-tracing method, by the total radiation from the emitter surface. The transport efficiency ratio is obtained by dividing the heat transport rate of the proposed emitter by

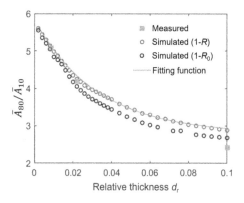

FIGURES 2.16 Experimental result of the ratio of average absorptivity, from 3 to 10 μm, between incident angles of $80°$ and $10°$ as a function of relative thickness defined as $d_r = d/(5 \text{ μm})$. The calculated $\bar{A}_{80}/\bar{A}_{10}$ from measured results are shown as square plots, and from simulated total absorptivity and corresponding fitting function are respectively in circle plots with a line. The black circle plots are the calculation from simulated normal reflectivity. (Reproduced from Ref. [20].)

TABLE 2.1 Analysis Results of the Radiative Heat Transport Rate with the Proposed Emitter and an Isotropic Emitter

h [cm]	This Emitter	Isotropic Emitter	Transport Efficiency Ratio
0.1	12.5%	0.2%	57.4
0.2	24.3%	0.9%	27.9
0.3	33.4%	1.9%	17.2
0.4	41.2%	3.4%	12.2
0.5	47.6%	5.2%	9.2
0.6	53.0%	7.3%	7.3
0.7	57.6%	9.6%	6.0
0.8	61.7%	12.2%	5.1
0.9	65.2%	14.9%	4.4
1	68.3%	17.7%	3.9

Source: This table is reproduced from supplementary of Ref. [20].

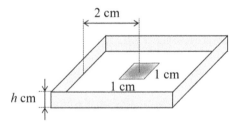

that of the isotropic emitter. For example, when we assume a 1 × 1 cm emission surface and surrounding walls that have a 0.5 cm height, which is 2 cm away from the emission surface center, only 5% of the emission to the total emission is reached in the case of the isotropic emission, whereas, 48% is reached using the proposed angular selective thermal emitter. The multiple bands throughout the IR range make it a flexible device for various applications at different temperatures. The peaks can be entirely characterized by the geometrical parameters, leading to flexible control of the thermal radiation. This addresses the material limitations often encountered with the angular selective emitter originated from BLM, which depends on the material intrinsic resonance frequencies.

2.4 CONCLUSION

We have developed a multiband and angular selective emitter (absorber) using the Berreman-leaky mode excited by coupling SPP into Fabry-Pérot-like resonances. This has been confirmed from exponential decays of the angular, selectivity depends on the relative thickness, both from the simulation and the experiment. When TM-polarized light is incidental from the grating direction perpendicular to the surface ($\theta = 85°$), the average absorptivity exceeds 0.5 in the range of 3–10 μm. By optimizing the structure, the peak angle's average absorptivity can be further increased since the absorption peaks are determined by the geometric parameters.

These technologies corresponding to the angular selective control of thermal radiation are expected to revolutionize thermal radiation management by offering unique characteristics such as flexible control over spectral properties and angular selectivity, distinct from those of natural thermal radiation.

REFERENCES

1. C. Zhang, C. Huang, M. Pu et al., "Dual-band wide-angle metamaterial perfect absorber based on the combination of localized surface plasmon resonance and Helmholtz resonance", *Sci. Rep.*, 7 (2017): 5652.
2. H. Zhu, F. Yi, E. Cubukcu, "Plasmonic metamaterial absorber for broadband manipulation of mechanical resonances", *Nat. Photonics*, 10 (2016): 709–714.
3. C. Argyropoulos, K. Q. Le, N. Mattiucci, G. D'Aguanno et al., "Broadband absorbers and selective emitters based on plasmonic Brewster metasurfaces", *Phys. Rev.*, 87 (2013): 205112 B.
4. C.-H. Fann, J. Zhang, M. E. Kabbash et al., "Broadband infrared plasmonic metamaterial absorber with multipronged absorption mechanisms", *Opt. Express*, 27 (2019): 27917.
5. H. Lin, B. C. P. Sturmberg, K. T. Lin et al., "A 90-nm-thick graphene metamaterial for strong and extremely broadband absorption of unpolarized light", *Nat. Photonics*, 13 (2019): 270.
6. T. Sang, J. Gao, X. Yin et al., "Angle-Insensitive Broadband Absorption Enhancement of Graphene Using a Multi-Grooved Metasurface", *Nanoscale Res. Lett.*, 14 (2019): 105.
7. L. P. Wang and Z. M. Zhang, "Resonance transmission or absorption in deep gratings explained by magnetic polaritons", *Appl. Phys. Lett*, 95 (2009): 111904.
8. Y.-H. Ye, Y.-W. Jiang, M.-W. Tsai et al., "Coupling of surface plasmons between two silver films in a Ag/SiO₂/Ag plasmonic thermal emitter with grating structure", *Appl. Phys. Lett.* 93 (2008): 263106.
9. S. Tsuda, S. Yamaguchi, Y. Kanamori et al., "Spectral and angular shaping of infrared radiation in a polymer resonator with molecular vibrational modes", *Opt. Express*, 26 (2018): 6899–6915.
10. T. Kosako, Y. Kadoya, and H. F. Hofmann. "Directional control of light by a nanooptical Yagi-Uda antenna", *Nat. Photonics*, 4 (2010): 312–315.
11. L. Zhu, F. Liu, H. Lin et al. "Angle-selective perfect absorption with two-dimensional materials" *Light Sci. Appl.*, 5 (2016): e16052–e16052.

12. N. C. Passler, I. Razdolski, D. S. Katzer et al. "Second harmonic generation from Phononic Epsilon-near-Zero Berreman modes in ultrathin polar crystal films", *ACS Photonics*, 6 (2019): 1365–1371.
13. S. Molesky. C. J. Dewalt, and Z. Jacob, "High temperature epsilon-near-zero and epsilon-nearpole metamaterial emitters for thermophotovoltaics", *Opt. Express*, 21 (2013): A96–A110.
14. S. Campione, F. Marquier, J. P. Hugonin et al., "Directional and monochromatic thermal emitter from epsilon-near-zero conditions in semiconductor hyperbolic metamaterials", *Sci. Rep.*, 6 (2016): 34746.
15. S. Vassant, J. P. Hugonin, F. Marquier et al. "Berreman mode and epsilon near zero mode", *Opt. Express*, 20 (2012): pp. 23971–23977.
16. W. D. Newman, C. L. Cortes, J. Atkinson et al. "Ferrell-berreman modes in plasmonic epsilon-near-zero media", *ACS Photonics*, 2 (2015): 2–7.
17. E. Sakr and P. Bermel. "Thermophotovoltaics with spectral and angular selective doped-oxide thermal emitters", *Opt. Express*, 25 (2017): A880.
18. I. Khan, Z. Fang, M. Palei et al. "Engineering the Berreman mode in mid-infrared polar materials", *Opt. Express*, 28 (2020): 28590.
19. J. Xu, J. Mandal, A. P. Raman, "Broadband directional control of thermal emission", *Science*, 372 (2021): 393.
20. R. Benlyas, M. Shimizu, K. Otomo et al., "Multiband and angular shaping of infrared radiation in 1D plasmonic gratings", *Opt. Express*, 30 (2022): 9380–9388.
21. R. A. Ferrell, "Predicted radiation of plasma oscillations in metal films", *Phys. Rev.*, 130 (1958): 1214–1222.
22. D. W. Berreman. "Infrared absorption at longitudinal optic frequency in cubic crystal films", *Phys. Rev.*, 130 (1963): 15.
23. K. Nisi, S. Subramanian, W. He et al., "Light–matter interaction in quantum confined 2D polar metals", *Adv. Funct. Mater.*, 31 (2021): 2005977.
24. G. Palermo, K. V. Sreekanth, N. Maccaferri et al., "Hyperbolic dispersion metasurfaces for molecular biosensing", *Nanophotonics*, 10 (2020): 295–314.
25. M. G. Moharam, T. K. Gaylord, "Rigorous coupled-wave analysis of planar-grating diffraction", *J. Opti. Soc. Ame.*, 71 (1981): 811–818.
26. P. Wu, C. Zhang, Y. Tang et al. "A perfect absorber based on similar fabry-perot four-band in the visible range", *Nanomaterials*, 10 (2020): 488.
27. H. Zhang, F. Ling, and B. Zhang. "Broadband tunable terahertz metamaterial absorber based on vanadium dioxide and Fabry-Perot cavity", *Opt. Mater.*, 112 (2021): 110803.
28. X. Li, Z. Zhu, Y. Xi et al., "Single-mode Fabry-Perot laser with deeply etched slanted double trenches", *Appl. Phys. Lett.*, 107 (2015): 091108.
29. X. Liu, H. Yang, X. Wang et al. "Hybrid Plasmonic Modes in Multilayer Trench Grating Structures", *Adv. Opt. Mat.*, 5 (2017): 1700496.
30. X. Liu, J. Gao, J. Gao et al., "Microcavity electrodynamics of hybrid surface plasmon polariton modes in high-quality multilayer trench gratings", *Light Sci. Appl.*, 7 (2018): 14.
31. M. K. Gunde "Vibrational modes in amorphous silicon dioxide", *Physica B: Condens. Matter*, 292 (2000): 286–295.

32. R. M. Nor, S. N. M. Halim, M. F. M. Taib et al., "First principles study on phonon energy in SiO$_2$ glass with the incorporation of Al$_2$O$_3$". *Solid State Phenomena 268 SSP* (2017):160–164.re
33. V. J. Gokhale, P. D. Myers, M. Rais-Zadeh, "Subwavelength plasmonic absorbers for spectrally selective resonant infrared detectors". *IEEE*, 24 (2014): 982–985.
34. J. Nie, J. Yu, W. Liu et al., "Ultra-narrowband perfect absorption of monolayer two-dimensional materials enabled by all-dielectric subwavelength gratings", *Opt. Express*, 28 (2020): 38592.
35. Z. Zhang, Z. Yu, Y. Liang et al., "Dual-band nearly perfect absorber at visible frequencies", *Opt. Mat. Express*, 8 (2018): 463.

CHAPTER 3

High Aspect Ratio Nanostructures for Photothermal Applications

Osamu Takayama

Technical University of Denmark, Kgs. Lyngby, Denmark

EFFICIENT USAGE OF ABUNDANT and free solar energy has immense impact towards the realization of a low carbon and sustainable society. In order to take advantage of solar energy, efficient conversion of light into a usable form of energy, such as electricity or heat, is of great importance. When it comes to the conversion of light to heat or heat to electromagnetic radiation for cooling, a material with high optical absorption or a large absorption cross-section is necessary. Apart from natural materials, artificially-engineered materials, often referred to as metamaterials or metasurfaces, may enable us to tailor the photothermal effect for a desired application. Recently, various nanostructured systems made of plasmonic (metallic) and dielectric materials have been studied. Among them, high aspect ratio nanostructures, whose height are a lot larger than the lateral periodicity, are shown to be effective in improving photothermal heating and radiative cooling. In this chapter, we provide an overview of current progress in photothermal applications on high aspect ratio nanostructures made of dielectric and plasmonic materials, especially titanium nitride.

DOI: 10.1201/9781003409090-3

3.1 INTRODUCTION

Heat is one of the most widely used forms of energy in everyday life from boiling water and cooking to the heating of living places. Among numerous ways to generate heat, exploitation of sunlight may be a more green source of energy for heat generation, rather than burning fossil fuels like petroleum. Photothermal effect is the conversion of energy from light to heat when light interacts with material and is absorbed by the material and turned into heat. There are numerous applications of photothermal effects that could contribute to a sustainable society. One of the potential applications is solar vapor or solar steam generation for desalination of undrinkable water into drinkable water by collecting the vapor [1], as illustrated in Figure 3.1(a). There is a growing demand for sufficient clean water for the population, which is a big global challenge due to the population growth [2], as currently, over 650 million people worldwide lack access to safe water [3]. Therefore, "water security" is an urgent issue for sustainable society. However, the conventional desalination process is, in principle, to burn precious petroleum to boil sea water and collect vapor as the drinkable water, which results in the generation of global warming gases and emission of pollutants. By taking advantage of broadband optically absorbing nanostructures, one can incorporate solar photothermal heating as a new approach to the distillation process.

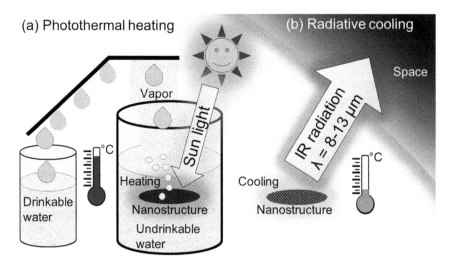

FIGURE 3.1 Conceptual illustration of potential applications of photothermal effect. (a) Heat generation for solar vapor to boil water by visible and near-infrared light (sun light) on nanostructures. (b) Radiative cooling.

Contrary to heating, efficient cooling also plays a significant role for energy saving in modern society. Today, cooling is required for living spaces, working places, storage areas, and machinery with elevated temperatures. Cooling is conventionally conducted by fans and air conditioners, which consume nearly 20 % of the electricity used in buildings around the globe today [4]. Recently, alternative cooling methods that do not require external energy consumption have been proposed and studied, namely *radiative cooling*. Radiative cooling is dumping the heat of objects into cold outer space at ~3 K in temperature in the form of electromagnetic radiation through the transparency window of earth's atmosphere that spans at 8–13 μm in wavelength, as illustrated in Figure 3.2. Heat of objects at a typical ambient temperature of ~300 K is radiated into the surroundings for mid-infrared (IR) wavelengths due to black-body radiation, determined by the temperature of the thermal emitter. For radiative cooling, absorptivity and emissivity of materials can be engineered and maximized for the atmospheric transparency window by nanostructuring, while maintaining low optical absorption in the visible and near-IR regions to avoid heating by sun light. Apart from these, there are numerous photothermal applications, such as photothermal imaging [5, 6], cancer therapy [6], chemical reactions and catalysts [7], and antibacterial surfaces [8].

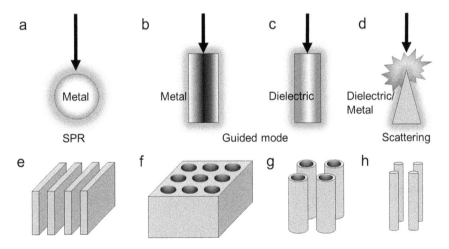

FIGURE 3.2 Conceptual illustration of high aspect ratio nanostructures and nano-optic phenomena. (a) Localized surface plasmon resonance (SPR) in plasmonic (metallic) nanoparticles, Propagating guided modes in (b) plasmonic and (c) dielectric nanostructures. (d) Scattering. (e) Trenches, (f) air hole, (g) tube, and (h) pillar arrays. The materials can be either plasmonic or dielectric.

Improved photothermal heating or cooling properties can be achieved by means of controlling the size and morphology of materials at nanoscale and surface texturing. These nanostructures may take the form of gratings, trenches, air hole arrays, pillars, and tubes, as depicted in Figure 3.2(a)–(d). There are several mechanisms for the enhancement of photothermal effect, that is, scattering, excitation of localized surface plasmon resonance (SPR) modes, propagating guided modes in dielectric or plasmonic materials, or a combination of these (Figure 3.2(e)–(h)). For instance, metallic nanostructures that support propagating or localized plasmon polariton modes result in high optical absorption and hence efficient generation of heat from light [9], although a bulk metal itself is very reflective and does not absorb light.

In this chapter, we discuss both dielectric and metallic nanostructures with high aspect ratio, their features, advantages, and related nano-optic phenomena behind photothermal applications mainly for solar heating and radiative cooling, as the design and realization of high aspect ratio nanostructures requires understanding of light-matter interaction at nanoscale. Throughout this chapter, we define "aspect ratio" as the ratio between the period of nanostructures in lateral dimension and height in vertical dimension. Moreover, we refer to "high" aspect ratio as an aspect ratio more than unity.

3.2 DIELECTRIC NANOSTRUCTURES

Dielectric materials whose real part of permittivities is positive exhibit less optical absorption in visible and near-IR wavelength regions as opposed to plasmonic material counterpart, which may not be suitable for photothermal applications. However by nanostructuring, light-matter interaction can be highly enhanced so that significant light can be absorbed in dielectric nanostructures for heating or radiating for the mid-IR wavelengths. In this section we discuss dielectric-based high aspect ratio nanostructures including silicon (Si), alumina (Al_2O_3), silica (SiO_2) and silicon carbide (SiC).

3.2.1 Black Silicon Nanostructures

Silicon is one of the most well-developed materials for nanofabrication thanks to microelectronics technology and its material abundance, which enables mass production and scalability, while reducing the cost. Doped Si pillar structures whose periods and height are on the order of a few to dozens of microns have been developed for solar cell applications [11].

For visible and near-IR wavelengths, Si is not as optically lossy as metals so it may not generate heat efficiently from sunlight. However, the surface of Si can be nanostructured to scatter light and couple to guided or localized modes. The key challenge is to design geometries of Si nanostructures where light is trapped within the Si nanostructures and absorbed by the material, generating heat from sunlight efficiently, and to realize nanofabrication processes.

Black Si is an anti-reflection Si surface with random roughness to reduce reflection, as opposed to flat Si surface, as illustrated in Figure 3.3(a) and (b). Black Si has been developed to improve light trapping for photovoltaic cells [12–14]. Depending on the fabrication methods and processes, black Si with various parameters can be fabricated [14]. For instance, by reactive ion etching, black Si with cone height of 222–1800 nm and cone diameter of 70–650 nm have been fabricated. The height of 100–4000 nm

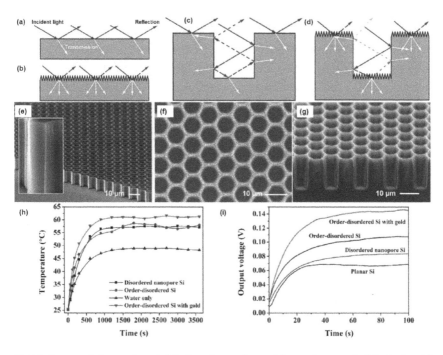

FIGURE 3.3 Schematic of (a) flat Si surface, (b) surface with black Si, (c) micropores, and (d) black Si micropores. Scanning electron microscopy (SEM) images of (e) bird's eye view of the porous Si (The inset shows the sidewall), (f) top view, and (g) cross-section view of black Si on the porous Si. (h) Water temperature and (i) the output voltages of the thermoelectric generator in terms of time. (Adapted with permission from Ref. [10]. Copyright 2019, Elsevier.)

and diameter of 25–130 nm have been produced by metal assisted chemical etching with metal nanoparticles as mask, while those of 550–2000 nm and 240–600 nm can be made by plasma immersion ion implantation.

High optical absorption can be achieved by the combination of high aspect ratio microstructures and black Si on microstructures that function as antireflection surface to trap light [10], as shown in Figure 3.3(c)–(g). The microstructures consist of air hole array in hexagonal lattice arrangement with 1 μm period and 2 μm height where top and bottom of the microstructure has black Si in order to increase the scattering and reduce reflection, leading to enhanced temperature increase by sun light [10]. Gold (Au) nanoparticles are distributed on the top surface by localized surface plasmon resonance in order to increase light absorption on the surface even further.

In order to characterize the photothermal effect of each structure immersed in water, the temperature rise of each structure was measured for the solar irradiation of 4 kW/m² (400 mW/cm²), as shown in Figure 3.3(h). The structure with the combination of microstructures, Black Si, and Au nanoparticles exhibit the highest temperature increase, as opposed to ones without Au nanoparticles. Not only direct generation of current by photovoltaic, the heat- generated light can be used to generate electric current via thermocouple or thermoelectric power generation module, called the thermophotovoltaic system. The nanostructures are placed on top of the thermoelectric generator module in order to characterize the generation of voltage by the temperature difference created by the nanostructures, as shown in Figure 3.3(i). The results show that the microstructured Si with anti-reflection and Au nanoparticles results in higher photothermal conversion and temperature, subsequently leading to higher voltage generated by thermoelectric generator. Additionally, similar hybrid systems with the combination of high aspect ratio dielectric structures and plasmonic materials have been studied. These examples include a square lattice air hole array in carbon nanotubes with pitch of 1800 nm as lossy dielectric and plasmonic ruthenium (Ru) nanoparticles to increase light absorption [15].

3.2.2 Highly-Ordered Silicon Nanostructures

Apart from black Si, high aspect ratio Si nanostructures can enhance photothermal heating properties by tuning the effective thermal conductivity through the ratio between Si and air holes or gaps [16]. Two types of Si nanostructures are considered here, air hole array and pillars in square

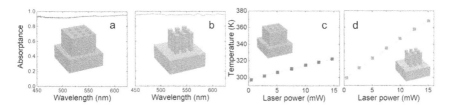

FIGURE 3.4   Absorptance of (a) air hole and (c) pillar array of Si. The photothermal heating temperature profile as a function of excitation laser power for (c) air hole and (d) pillar array. The depth of air holes and the height of Si pillars was 3.2 μm and the period of 0.6 μm for both structures. Air hole array has a hole/period ratio of 0.76. Pillar array has a pillar/period ratio of 0.54. (Adapted with permission from ref. [16]. Copyright 2023, Elsevier.)

lattice arrangement, as shown in Figure 3.4. These Si nanostructures can be fabricated by either deep ultraviolet (UV) or electron beam lithography to pattern Si wafer and subsequent deep reactive ion etching, a dry etching technique, to etch Si wafers into the designed nanostructures. The air hole and pillar arrays have been fabricated with height range of 0.32–5.7 μm, periods of 0.4–1.0 μm, which gives an aspect ratio of ~10. The hole/period and pillar/period ratio range from 0.59 to 0.85 and 0.27 to 0.68, respectively.

Both structures exhibit over 90 % (0.9) absorptance in the visible wavelength range (Figure 3.4(a) and (b)). Temperature increase by laser irradiation at 514 nm in terms of power was measured via shift of the Raman spectra of Si, which appears at 520.6 cm^{-1} and shifts for different temperatures. Prior to the measurement, a calibration curve relating the temperature of Si and Raman spectra was made by recording a series of Stokes peak shifts on a bare Si substrate with a heating stage that controls the temperature. In fabricated Si nanostructures, heat is generated as light propagates along Si pillars and matrix and is absorbed by Si. Photothermal heating was observed for increasing laser power, as shown in Figure 3.4(c) and (d), and a linear dependence can be seen within the laser power range of 0–15 mW. The temperature increase is higher for Si pillars than air hole arrays in Si matrix. This is be because the temperature rise is not only influenced by the generated heat in the structures by light absorption but also strongly influenced by the effective thermal conductivity of the structure. Si pillar arrays show a larger photothermal heating effect than the air hole arrays, since Si pillars are separated by air gaps with very low thermal conductivity, leading to lower in-plane thermal conductivity and, consequently, high temperature rise. On the other hand, an air hole array in square lattice arrangement has inter-connection of unit cell by Si with high thermal conductivity, so they exhibit relatively higher effective thermal conductivity in a lateral plane. In this case, generated heat can escape in the lateral plane, as well as vertical direction to the substrate, resulting in a lower temperature increase. The effective thermal conductivity can be tailored by varying their geometrical parameters, such as air hole and pillar diameters as opposed to period.

The heating capability can also be controlled by the pillar height, as taller pillars result in higher surface temperatures. This is because heat dissipation to substrate along the vertical direction is also a crucial issue for nanostructures on substrates with high thermal conductivity, such as silicon and sapphire where generated heat escapes into the substrate and the temperature does not increase [17]. High aspect ratio nanostructures are also advantageous in that undesirable heat dissipation into the substrate is reduced. Even a height of a few to several micrometers of those structures is sufficient to suppress the heat conduction to the substrates and subsequently enhance the photothermal heating by the incident light and may significantly contribute to photothermal heating applications.

3.2.3 Anodic Porous Alumina

As opposed to top-down approaches to fabricate nanostructures from bulk materials by patterning and etching, there are bottom-up approaches to generate nanostructures via chemical process. One of such self-assembled nanostructures, called either anodic porous alumina (APA) or anodic aluminum oxide (AAO), is an example of high aspect ratio nanostructures. APA can be produced by anodization of aluminum (Al) film or foil [18] (Figure 3.5(a) and (b)). APA nanostructures have a height of 50 and 100 μm, as opposed to the period of 50–500 nm, making the aspect ratio extremely high, ranging from dozens to hundreds. APA has been studied for various optical applications, such as photonic crystals [19–21] and templates for producing metallic nanowires, including Ni, Au [22].

APA has also been studied as a platform for daytime passive radiative cooling. For passive radiative cooling, ideally the material should possess high absorption and emittance for the atmospheric transparency windows of 8–13 μm in wavelength and be transparent or reflective for the solar spectrum region. APA's effective permittivities and thermal conductivities can be tuned by air hole sizes in order to overlap its absorptive region with the transparency window [25, 26]. An APA structure was fabricated to

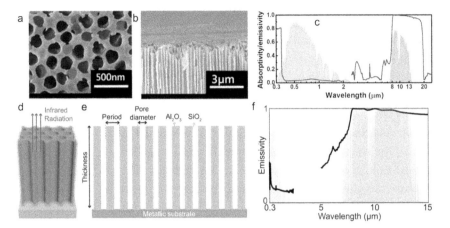

FIGURE 3.5   SEM images of (a) top and (b) cross section view of APA sample. (c) Solar spectrum and atmospheric window (shaded region) in a relative humidity of 70 %, and 70 % and measured spectral absorptivity and emissivity of APA sample. Adapted with permission from ref. [23]. Copyright 2019, Elsevier. (d), (e) Schematic illustration of APA coated with 15 nm thick SiO_2 film for radiative cooling. APA has the period of 500 nm, pore diameter of 400 nm, and thickness of 50 μm. (f) Measured emissivity spectrum of the APA structures from UV to mid-IR region. (Adapted with permission from Ref. [24]. Copyright 2021, Elsevier.)

generate elevated absorptivity and emissivity at 8–20 μm, overlapping with the atmosphere's IR transmission window of 8–13 μm, while for other wavelengths absorptivity and emissivity are kept low [23], as shown in Figure 3.5(c). The APA structure had a period of around 340 nm with volume faction of air holes 0.82 and height of 56.8 μm on aluminum (Al) substrate. The fabricated APA nanostructures exhibited a potential cooling power density of 64 W/m² (6.4 mW/cm²) at ambient temperature (humidity ~70 %) under sunlight irradiance and a 2.6 K temperature reduction below the ambient air temperature.

Later, the combination of APA and silica layer was studied to improve cooling capability [24]. The APA structure had 500 nm period, 400 nm air hole diameter, and 15 nm thick silica layer conformally deposited by the atomic layer deposition (ALD) technique, as illustrated in Figure 3.5(d) and (e). Bare alumina has absorption above 10 μm in wavelength due to its phonon, while absorption from 8 μm in wavelength is desirable for radiative cooling. Silica has phonon peaks for 8–10 μm, adding an extra absorption band to cover the transparency window, as shown in Figure 3.5(f). With the composite APA nanostructures, maximum cooling of 6.1 K and cooling power density of 65.6 W/m² (6.56 mW/cm²) were achieved.

3.2.4 Silica and Silicon Carbide

Materials used for radiative cooling need to be transparent in the solar spectral region and produce no heat, meanwhile they are required to exhibit high absorption over the atmospheric transparency window in the mid-IR region. Silica is transparent in the visible and near-IR regions and its absorption in the atmospheric transparency region can be increased by nanostructuring, while maintaining the transparency in the solar spectral region. One such structure has been realized by a photonic crystal in silica with air hole arrays in square lattice arrangement. The photonic crystal structure has the period of 6 µm and the depth of air holes 10 µm fabricated by photolithography on 500 µm thick glass substrate. The role of the photonic crystal is to couple incident light to the phonon mode of silica, increasing the absorption and emission in the atmospheric transparency windows of 8–13 µm and resulting in 13 K cooling from the temperature of 525 µm thick Si wafer as a reference.

Furthermore, the combination of silica pillars and poly(methyl methacrylate) (PMMA) polymer was proposed to maintain high optical transparency in the visible wavelength while maintaining high absorption in the mid-IR region. The silica pillars fabricated on fused silica substrate were 1250 nm tall, with 172–307 nm diameter and average period of 410 nm, embedded in a 1500 nm thick PMMA layer [28]. Silica pillars increased absorption in the mid-IR wavelength region by scattering and coupling to surface phonon polariton modes, but also made the fused silica hazy and less transparent in the visible region. Coating the fused silica surface by PMMA layer with a similar refractive index as silica improved the transparency for visible wavelengths by making the top surface flat. The silica pillar structures have been fabricated by the combination of sputtering, reactive ion etching, and subsequent spin coating of PMMA. The nanostructures showed 16 K lower than the bare fused glass substrate at 407 K, while maintaining high transparency in the visible wavelength, suitable for optical displays applications.

Apart from silica, silicon carbide (SiC)-based nanostructures have also been studied for radiative cooling, as shown in Figure 3.6 [27]. SiC is a transparent polar material in the solar spectrum but possesses the Reststrahlen band or phonon resonance band at 10.5–12.6 µm where its real part of permittivity becomes negative and reflects light significantly, making high absorption in the atmospheric transparency window challenging. Therefore, in order to ameliorate the difficulty, SiC pillars with

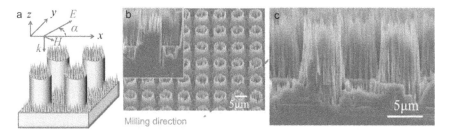

FIGURE 3.6   (a) Illustration of SiC structures with macro pillars with nanocones. SEM images of (b) bird's eye view of top and cross-section views. The inset is the cross section of a pillar. (c) SEM image of the cross section of the fabricated SiC pillars with cones made by milling the sample along the direction shown in (b) (dashed line). (Adapted with permission from ref. [27]. Copyright 2018, Royal Society of Chemistry.)

period of 8 µm, diameter 6 µm, and 8.3 µm with cone-like surface has been investigated. The periodic micro pillars as a grating coupler achieve near perfect absorption at part of SiC's Reststrahlen band. In order to achieve high absorption in the broad Reststrahlen band, the cones scatter mid-IR light with broadband and couple to bulk and surface phonon polariton modes, which are eventually absorbed. As a result, the SiC structure achieved high absorption of over 97 % in the Reststrahlen band.

3.3 PLASMONIC NANOSTRUCTURES – TITANIUM NITRIDE

Plasmonics is a sub-field of nanophotonics where propagating or localized surface plasmon polaritons play essential role to confine light at nanoscale on metallic nanostructures whose permittivities are negative. In general, the large optical absorption or loss of plasmonic materials is one of the biggest challenges for the exploitation of plasmonic nanostructures for practical applications. On the other hand, photothermal applications take the advantage of large light absorption and absorption cross section of plasmonic nanostructures for efficient light to heat conversion. In plasmonic photothermal heating, the heated region can be confined even to the subwavelength scale [5].

Recently metallic nanoparticles have been studied to enhance light absorption and boil water for water purification purpose [29–31], where the heating materials composed of nanostructures are immersed in liquid [17]. A key requirement here is the ability to absorb light in the solar spectrum that spans over the visible and infrared wavelengths by multiple scattering [30]. Gold (Au) has been exclusively used for photothermal heating owing to its well-known plasmonic properties and chemical stability [9], though its melting temperature, ~1000°C, may be a limiting factor for high temperature thermal applications.

Apart from Au-based nanostructures, recently, titanium nitride (TiN) [32] as an alternative plasmonic material has been studied for photothermal nanostructures because the optical losses of TiN are moderately larger than those of Au, so it is advantageous for obtaining elevated light absorption by plasmon resonances. TiN is a ceramic and a transition metal nitride with a higher melting temperature, ~2000°C, which is favourable especially for high temperature operation of photothermal applications. Active research on plasmonic photothermal heating with TiN has been carried out during the past decade in terms of local heating by nanodiscs or particles [17, 33, 34], solar heating and distillation [35], and catalysts to promote chemical reactions [7]. These structures have the heights of a few tens of nanometers and the generated heat by the nanostructures dissipates into the substrates before high temperatures are realized in the irradiated area. The leakage of heat to substrate can be suppressed by a plasmonic photothermal nanostructure with a high optical absorptivity and low thermal conductance.

Similar to previously discussed Si nanostructures, high-aspect-ratio TiN trench and tube structures were demonstrated for photothermal applications, as shown in Figure 3.7(a) and (b) [36]. Such high aspect ratio trench, pillar, and tube structures can be fabricated by the combination of deep UV lithography for patterning, dry etching technique, and atomic layer deposition (ALD) of TiN [36, 37]. This scheme enables us to fabricate high aspect ratio structures with different materials, such as Al_2O_3, titania (TiO_2) [38, 39], zinc oxide (ZnO)-based [40–44] nanostructures. The fabricated TiN tube structures are 400 nm period, ~300 nm diameter, 2–3 μm height for nanotubes. For trench structures, 200 nm air gap and wall thickness are typically made with 400 nm period. The temperature increase of each structure was measured by monitoring the Raman shift similar to the case of Si nanostructures. Prior to temperature the measurement of TiN nanostructures, Raman spectrum of TiN film for different temperature was characterized in order to obtain the calibration curves, which enables us to monitor the temperature during photothermal heating for different laser irradiation power at 785 nm in wavelength.

Among the tubes and trenches, the tube structures have superior photothermal properties and temperature rise. The enhanced photothermal

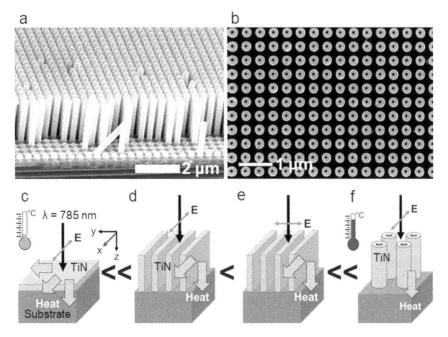

FIGURE 3.7 SEM images of TiN tubes in (a) birds-eye-view, and (b) top view. Conceptual illustration of isotropic and anisotropic heat propagation for each structures. (c) TiN film on Si substrate. (d) TiN trenches with electric field (polarization) parallel and (e) perpendicular to trenches on Si substrate. (f) TiN tubes on Si substrate. Temperature increase by laser irradiation at 785 nm is larger in the order of tubes, trenches with perpendicular polarization, parallel polarization, and films on Si substrate.

heating is attributed to the extremely anisotropic effective thermal conductivities according to the finite element heat transfer analyses. In the case of TiN film deposited on Si substrate (Figure 3.7(c)), heat can escape toward all three x, y, and z directions, which makes it difficult to obtain high temperature for laser irradiation. For TiN trenches, one out of the three components of the anisotropic thermal conductivity (along y-axis in Figure 3.7(d) and (e)) has three orders of magnitude smaller than the bulk thermal conductivity of TiN due to the air gap between TiN trenches. Note that the temperature rise of TiN trenches parallel to polarization is lower than those of perpendicular polarization because incident light with parallel polarization is reflected more and results in lower absorption. In the case of tubes, two out of three thermal conductivity components (along x- and y-axis in Figure 3.7(f)) are low due to air gaps between tubes. As a result, these high aspect ratio TiN nanotube structures showed a 100-fold enhancement of photothermal temperature increase compared to the planar TiN film. When engineering the effective thermal conductivity of nanostructures, the geometries play a major role in reducing the heat conduction. Trench and pillar (including tube) arrays are disconnected in lateral directions, which are considered to be effective in lowering the effective thermal conductivities in plane.

High aspect ratio APA structures can be used as templates of plasmonic nanostructures. TiN film was deposited by sputtering on top of APA where TiN were partially infiltrated in air hole array, which was studied for solar vapor production, as shown in Figure 3.8 [35]. The fabricated APA nanostructures had the height of 50 and 100 μm, as opposed to the period of 150–500 nm, making the aspect ratio extremely high, ranging from 200 to 333. It turned out that the structure with period 400 nm and hole diameter of 300 nm showed the larger evaporation of water among other TiN-coated APA nanostructures, and thus the better desalination capability. Moreover, APA nanostructures were coated with 80 nm thick TiN film, whose heat transfer and photothermal properties were experimentally and theoretically analyzed [45]. The diameter of air holes were varied from 161 to 239 nm, depending on anodization time. Highest optical absorptance and photothermal heating efficiency are observed for the air hole diameter of 239 nm. Apart from TiN, the combination of APA and ziconium nitride, hafnium nitride [46] and aluminum nanoparticles [47] has been investigated for water desalination.

3.4 CONCLUSION

In summary, we discuss high aspect ratio nanostructures made of either plasmonic or dielectric elements, including titanium nitride, silicon, alumina, and silica. The underlying nano-optic phenomena supported in these nanostructures vary from localized plasmon resonance in metallic nanostructures to guided modes in dielectric nanostructures. In general, plasmonic nanostructures are able to generate more intense heat and result in higher temperature rise than dielectric counterpart. Finally, Table 3.1 summarizes the nanostructures discussed in this chapter. Adopting a high aspect ratio nanostructure to realize low thermal conductivity is expected to work for other materials. Using low thermal conductivity materials, such as silica and zirconia for nanostructures may result in even lower thermal conductivity than the current samples. For passive radiative cooling, a transparent material for solar spectrum with phonon absorption for

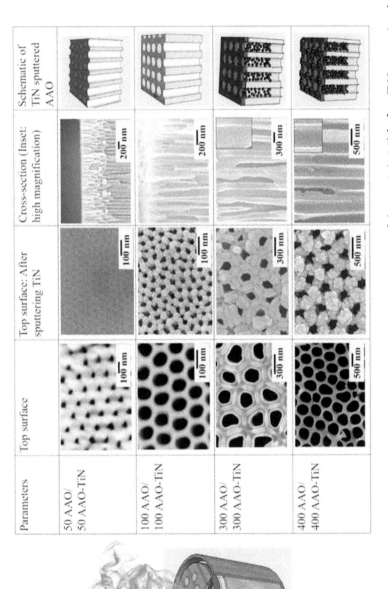

FIGURE 3.8 Conceptual image of the steam generation by TiN on APA. SEM images of the AAO (APA) before TiN sputtering (top view), after sputtering TiN on AAO (top view and cross-sectional view), and sketches ofc AAO – TiN. (Adapted with permission from ref. [35]. Copyright 2018, John Wiley & Sons, Inc.)

TABLE 3.1   Examples of High Aspect Ratio Nanostructures for Photothermal Applications

Structure	Material	Operating Wavelength [nm]	Power [mW]	Temperature Increase/ Decrease [K]	Ref.
Dielectric	material				
Black Si	Si + Au NPs	200–1700	400 mW/cm²	RT → 335	[10]
Pillars, hole array	Si	514	0 – 15	RT → 400	[16]
Photonic	material	For	radiative	cooling	
Air hole array PC	SiO₂	0.3–1.8 μm, 6–25 μm	–	−13	[48]
Pillars	SiC	10.5–12.6 μm	–	–	[27]
Air hole array	APA	0.3–2.5 μm, 2.5 – 20 μm	*6.4 mW/cm²	−2.6	[23]
Air hole array	APA + SiO₂	0.3–2 μm, 5–15 μm	*6.56 mW/cm²	−6.1	[24]
Pillars	SiO₂ + PMMA	0.3–1.5 μm, 7–11 μm	*14 mW/cm²	−16	[28]
Plasmonic	material				
Trenches, tubes	TiN	785	0–30	RT → 800	[36]
Pillar array and film	TiN + APA	Solar irradiation	100, 800 mW/cm²	RT → 373	[35]
Hole array and film	TiN + APA	785	0–5	RT → 410	[45]
Air hole array	Al + APA	Solar irradiation	100–600 mW/cm²	RT → 373	[47]
Air hole array	TiN, ZrN, HfN+APA	Solar irradiation	100–400 mW/cm²	RT → 323	[46]

APA: anodic porous alumina, NP: nanoparticle, PC: photonic crystal, PMMA: poly(methyl methacrylate), RT: room temperature is around ~293 K. However, it varies, depending on the research. "Solar irradiation" comes from a solar simulator whose emission resembles that of the solar spectrum. *Cooling power density.

atmospheric transparency windows in mid-IR wavelength can be used by nanostructuring. The understanding of the relationship between nanostructure dimensions and their photothermal properties may guide the future design of novel nanostructures for photothermal applications.

REFERENCES

[1] Hadi Ghasemi, George Ni, Amy Marie Marconnet, James Loomis, Selcuk Yerci, Nenad Miljkovic, and Gang Chen. Solar steam generation by heat localization. *Nature Communications*, Vol. 5, p. 4449, 2014.

[2] B. Chen, M. Y. Han, K. Peng, S. L. Zhou, L. Shao, X. F. Wu, W. D. Wei, S. Y. Liu, Z. Li, J. S. Li, and G. Q. Chen. Global land-water nexus: Agricultural land and freshwater use embodied in worldwide supply chains. *Science of the Total Environment*, Vol. 613–614, pp. 931–943, 2018.

[3] UNICEF & WHO. Progress on Sanitation and Drinking Water: 2015 Update and MDG Assessment. *UNICEF*, 2015.

[4] Fatih Birol. The Future of Cooling. *International Energy Agency*, 2018.

[5] Guillaume Baffou, Christian Girard, and Romain Quidant. Mapping heat origin in plasmonic structures. *Physical Review Letters*, Vol. 104, No. 13, p. 136805, 2010.

[6] Wenya He, Kelong Ai, Chunhuan Jiang, Yuanyuan Li, Xiangfu Song, and Lehui Lu. Plasmonic titanium nitride nanoparticles for in vivo photoacoustic tomography imaging and photothermal cancer therapy. *Biomaterials*, Vol. 132, pp. 37–47, 2017.

[7] Alberto Naldoni, Zhaxylyk A. Kudyshev, Luca Mascaretti, Smritakshi P. Sarmah, Sourav Rej, Jens P. Froning, Ondřej Tomanec, Jeong Eun Yoo, Di Wang, Štěpán Kment, Tiziano Montini, Paolo Fornasiero, Vladimir M. Shalaev, Patrik Schmuki, Alexandra Boltasseva, and Radek Zbořil. Solar thermoplasmonic nanofurnace for high-temperature heterogeneous catalysis. *Nano Letters*, Vol. 20, No. 5, pp. 3663–3672, 2020.

[8] Jing Wei Xu, Ke Yao, and Zhi Kang Xu. Nanomaterials with a photothermal effect for antibacterial activities: An overview. *Nanoscale*, Vol. 11, No. 18, pp. 8680–8691, 2019.

[9] Guillaume Baffou and Romain Quidant. Thermo-plasmonics: Using metallic nanostructures as nano-sources of heat. *Laser and Photonics Reviews*, Vol. 7, No. 2, pp. 171–187, 2013.

[10] Zengxing Zhang, Yonghua Wang, Per Anders Stensby Hansen, Kang Du, Kim Robert Gustavsen, Guohua Liu, F. Karlsen, Ola Nilsen, Chenyang Xue, and Kaiying Wang. Black silicon with order-disordered structures for enhanced light trapping and photothermic conversion. *Nano Energy*, Vol. 65, No. June, p. 103992, 2019.

[11] Rick Elbersen, Wouter Vijselaar, Roald M. Tiggelaar, Han Gardeniers, and Jurriaan Huskens. Fabrication and Doping Methods for Silicon Nano- and Micropillar Arrays for Solar-Cell Applications: A Review. *Advanced Materials*, Vol. 27, No. 43, pp. 6781–6796, 2015.

[12] Xiaogang Liu, Paul R. Coxon, Marius Peters, Bram Hoex, Jacqueline M. Cole, and Derek J. Fray. Black silicon: Fabrication methods, properties and solar energy applications. *Energy and Environmental Science*, Vol. 7, No. 10, pp. 3223–3263, 2014.

[13] Qiulin Tan, Fengxiang Lu, Chenyang Xue, Wendong Zhang, Liwei Lin, and Jijun Xiong. Nano-fabrication methods and novel applications of black silicon. *Sensors and Actuators, A: Physical*, Vol. 295, pp. 560–573, 2019.

[14] J. Y.H. Chai, B. T. Wong, and S. Juodkazis. Black-silicon-assisted photovoltaic cells for better conversion efficiencies: a review on recent research and development efforts. *Materials Today Energy*, Vol. 18, p. 100539, 2020.

[15] Zhequn Huang, Changhong Cao, Qixiang Wang, Heng Zhang, Crystal Elaine Owens, A John Hart, and Kehang Cui. Multiscale plasmonic refractory nanocomposites for high-temperature solar photothermal conversion. *Nano Letters*, Vol. 22, pp. 8526–8533, 2022.

[16] Satoshi Ishii, Nicholaus K Tanjaya, Evgeniy Shkondin, Shunsuke Murai, and Osamu Takayama. Materials Today Nano Optical absorption and heat conduction control in high aspect ratio silicon nanostructures for photothermal heating applications Optical absorption and heat conduction control in high aspect ratio silicon nanostructures for phototherm. *Applied Materials Today*, Vol. 32, p. 101824, 2023.

[17] Satoshi Ishii, Ryosuke Kamakura, Hiroyuki Sakamoto, Thang Duy Dao, Satish Laxman Shinde, Tadaaki NAGAO, Koji Fujita, Kyoko Namura, Motofumi Suzuki, Shunsuke Murai, and Katsuhisa Tanaka. Demonstration of temperature-plateau superheated liquid by photothermal conversion of plasmonic titanium nitride nanostructures. *Nanoscale*, Vol. 10, pp. 18451–18456, 2018.

[18] Hideki Masuda, Haruki Yamada, Masahiro Satoh, Hidetaka Asoh, Masashi Nakao, and Toshiaki Tamamura. Highly ordered nanochannel-array architecture in anodic alumina. *Applied Physics Letters*, Vol. 71, No. 19, pp. 2770–2772, 1997.

[19] Hideki Masuda, Masayuki Ohya, Hidetaka Asoh, Masashi Nakao, Masaya Nohtomi, and Toshiaki Tamamura. Photonic crystal using anodic porous alumina. *Japanese Journal of Applied Physics, Part 2: Letters*, Vol. 38, No. 12 A, pp. L1403–L1405, 1999.

[20] Jinsub Choi, Yun Luo, Ralf B. Wehrspohn, Rainald Hillebrand, Jörg Schilling, and Ulrich Gösele. Perfect two-dimensional porous alumina photonic crystals with duplex oxide layers. *Journal of Applied Physics*, Vol. 94, No. 8, pp. 4757–4762, 2003.

[21] Osamu Takayama and Michael Cada. Two-dimensional metallo-dielectric photonic crystals embedded in anodic porous alumina for optical wavelengths. *Applied Physics Letters*, Vol. 85, No. 8, pp. 1311–1313, 2004.

[22] X. Y. Zhang, L. D. Zhang, W. Chen, G. W. Meng, M. J. Zheng, L. X. Zhao, and F. Phillipp. Electrochemical fabrication of highly ordered semiconductor and metallic nanowire arrays. *Chemistry of Materials*, Vol. 13, No. 8, pp. 2511–2515, 2001.

[23] Y. Fu, J. Yang, Y. S. Su, W. Du, and Y. G. Ma. Daytime passive radiative cooler using porous alumina. *Solar Energy Materials and Solar Cells*, Vol. 191, No. October 2018, pp. 50–54, 2019.

[24] Dasol Lee, Myeongcheol Go, Soomin Son, Minkyung Kim, Trevon Badloe, Heon Lee, Jin Kon, and Junsuk Rho. Sub-ambient daytime radiative cooling by silica-coated porous anodic aluminum oxide. *Nano Energy*, Vol. 79, p. 105426, 2021.

[25] D. A. Borca-Tasciuc and G. Chen. Anisotropic thermal properties of nano-channeled alumina templates. *Journal of Applied Physics*, Vol. 97, No. 8, p. 084303, 2005.

[26] Begoña Abad, Jon Maiz, Alejandra Ruiz-Clavijo, Olga Caballero-Calero, and Marisol Martin-Gonzalez. Tailoring thermal conductivity via three-dimensional porous alumina. *Scientific Reports*, Vol. 6, p. 38595, 2016.

[27] Dongxue Chen, Jianjie Dong, Jianji Yang, Yilei Hua, Guixin Li, Chuanfei Guo, Changqing Xie, Ming Liu, and Qian Liu. Realization of near-perfect absorption in the whole reststrahlen band of SiC. *Nanoscale*, Vol. 10, No. 20, pp. 9450–9454, 2018.

[28] Javier Arrés Chillón, Bruno Paulillo, Prantik Mazumder, and Valerio Pruneri. Transparent Glass Surfaces with Silica Nanopillars for Radiative Cooling. *ACS Applied Nano Materials*, Vol. 5, pp. 17606–17612, 2022.

[29] Oara Neumann, Alexander S. Urban, Jared Day, Surbhi Lal, Peter Nordlander, and Naomi J. Halas. Solar vapor generation enabled by nanoparticles. *ACS Nano*, Vol. 7, No. 1, pp. 42–49, 2013.

[30] Nathaniel J. Hogan, Alexander S. Urban, Ciceron Ayala-Orozco, Alberto Pimpinelli, Peter Nordlander, and Naomi J. Halas. Nanoparticles heat through light localization. *Nano Letters*, Vol. 14, No. 8, pp. 4640–4645, 2014.

[31] Pratiksha D. Dongare, Alessandro Alabastri, Seth Pedersen, Katherine R. Zodrow, Nathaniel J. Hogan, Oara Neumann, Jinjian Wu, Tianxiao Wang, Akshay Deshmukh, Menachem Elimelech, Qilin Li, Peter Nordlander, and Naomi J. Halas. Nanophotonics-enabled solar membrane distillation for off-grid water purification. *Proceedings of the National Academy of Sciences*, Vol. 114, No. 27, pp. 6936–6941, 2017.

[32] Matthew P. Wells, Ryan Bower, Rebecca Kilmurray, Bin Zou, Andrei P. Mihai, Gomathi Gobalakrichenane, Neil McN. Alford, Rupert F. M. Oulton, Lesley F. Cohen, Stefan A. Maier, Anatoly V. Zayats, and Peter K. Petrov. Temperature stability of thin film refractory plasmonic materials. *Optics Express*, Vol. 26, No. 12, p. 15726, 2018.

[33] Urcan Guler, Justus C. Ndukaife, Gururaj V. Naik, A. G. Agwu Nnanna, Alexander V. Kildishev, Vladimir M. Shalaev, and Alexandra Boltasseva. Local heating with titanium nitride nanoparticles. *Nano Lett.*, Vol. 13, pp. 6078–6083, 2013.

[34] Satoshi Ishii, Ramu Pasupathi Sugavaneshwar, and Tadaaki Nagao. Titanium nitride nanoparticles as plasmonic solar heat transducers. *Journal of Physical Chemistry C*, Vol. 120, No. 4, pp. 2343–2348, 2016.

[35] Manpreet Kaur, Satoshi Ishii, Satish Laxman Shinde, and Tadaaki Nagao. All-ceramic solar-driven water purifier based on anodized aluminum oxide and plasmonic titanium nitride. *Advanced Sustainable Systems*, Vol. 3, p. 1800112, 2018.

[36] Satoshi Ishii, Makoto Higashino, Shinya Goya, Evgeniy Shkondin, Katsuhisa Tanaka, Tadaaki Nagao, Osamu Takayama, and Shunsuke Murai. Extreme thermal anisotropy in high-aspect-ratio titanium nitride nanostructures for efficient photothermal heating. *Nanophotonics*, Vol. 10, No. 5, pp. 1487–1494, 2021.

[37] E. Shkondin, T. Repän, O. Takayama, and A. V. Lavrinenko. High aspect ratio titanium nitride trench structures as plasmonic biosensor. *Optical Materials Express*, Vol. 7, No. 11, pp. 4171–4182, 2017.

[38] Evgeniy Shkondin, Osamu Takayama, Jonas Michael Lindhard, Pernille Voss Larsen, Mikkel Dysseholm Mar, Flemming Jensen, and Andrei V. Lavrinenko. Fabrication of high aspect ratio TiO2 and Al2O3 nanogratings by atomic layer deposition. *Journal of Vacuum Science & Technology A: Vacuum, Surfaces, and Films*, Vol. 34, No. 3, p. 031605, 2016.

[39] Evgeniy Shkondin, Hossein Alimadadi, Osamu Takayama, Flemming Jensen, and Andrei V. Lavrinenko. Fabrication of hollow coaxial Al 2 O 3 / ZnAl 2 O 4 high aspect ratio freestanding nanotubes based on the Kirkendall effect. *Journal of Vacuum Science & Technology A: Vacuum, Surfaces, and Films*, Vol. 38, p. 013402, 2020.

[40] Osamu Takayama, Evgeniy Shkondin, Andrey Bogdanov, Mohammad Esmail Aryaee Pahah, Kirill Golenitskii, Pavel A. Dmitriev, Taavi Repän, Radu Malreanu, Pavel Belov, Flemming Jensen, and Andrei V. Lavrinenko. Midinfrared surface waves on a high aspect ratio nanotrench platform. *ACS Photonics*, Vol. 4, No. 11, pp. 2899–2907, 2017.

[41] Osamu Takayama, Pavel Dmitriev, Evgeniy Shkondin, Oleh Yermakov, Mohammad Esmail Aryaee Panah, Kirill Golenitskii, Flemming Jensen, Andrey Bogdanov, and Andrei V. Lavrinenko. Experimental observation of Dyakonov Plasmons. *Semiconductors*, Vol. 52, No. 4, pp. 442–446, 2018.

[42] Evgeniy Shkondin, Taavi Repän, Mohammad Esmail Aryaee Panah, Andrei V. Lavrinenko, and Osamu Takayama. High aspect ratio plasmonic nanotrench structures with large active surface area for label-free mid-infrared molecular absorption sensing. *ACS Applied Nano Materials*, Vol. 1, No. 3, pp. 1212–1218, 2018.

[43] Sharmistha Chatterjee, Evgeniy Shkondin, Osamu Takayama, Adam Fisher, Arwa Fraiwan, Umut A. Gurkan, Andrei V. Lavrinenko, and Giuseppe Strangi. Hydrogen gas sensing using aluminum doped ZnO metasurface. *Nanoscale Advances*, Vol. 2, pp. 3452–3459, 2020.

[44] Sharmistha Chatterjee, Evgeniy Shkondin, Osamu Takayama, Andrei V. Lavrinenko, Michael Hinczewski, and Giuseppe Strangi. Generalized Brewster effect in aluminum-doped ZnO nanopillars. *Proceedings of SPIE*, Vol. 1134524, No. April, p. 72, 2020.

[45] Nicholaus Kevin Tanjaya, Manpreet Kaur, Tadaaki Nagao, and Satoshi Ishii. Photothermal heating and heat transfer analysis of anodic aluminum oxide with high optical absorptance. *Nanophotonics*, Vol. 11, No. 14, pp. 3375–3381, 2022.

[46] Emily Traver, Reem A. Karaballi, Yashar E. Monfared, Heather Daurie, Graham A. Gagnon, and Mita Dasog. TiN, ZrN, and HfN nanoparticles on nanoporous aluminum oxide membranes for solar-driven water evaporation and desalination. *ACS Applied Nano Materials*, Vol. 3, No. 3, pp. 2787–2794, 2020.

[47] Lin Zhou, Yingling Tan, Jingyang Wang, Weichao Xu, Ye Yuan, Wenshan Cai, Shining Zhu, and Jia Zhu. 3D self-assembly of aluminium nanoparticles for plasmon-enhanced solar desalination. *Nature Photonics*, Vol. 10, No. 6, pp. 393–398, 2016.

[48] Linxiao Zhu, Aaswath P. Raman, and Shanhui Fan. Radiative cooling of solar absorbers using a visibly transparent photonic crystal thermal blackbody. *Proceedings of the National Academy of Sciences of the United States of America*, Vol. 112, No. 40, pp. 12282–12287, 2015.

CHAPTER 4

Thermal High-Contrast Metamaterials

Richard Z. Zhang
University of North Texas, Denton, TX

Ken Araki
Arizona State University, Tempe, AZ

4.1 INTRODUCTION

This chapter is dedicated to the design and implementation of advanced thermal radiative coatings and surface treatments. These thermal radiative materials can lead to a low-carbon future as illustrated in Figure 4.1, by improving the energy efficiency of buildings, vehicles, and industrial processes, enhancing reliability and performance of electronic devices, reducing reliance on active cooling systems, and uncovering novel methods of renewable energy harvesting and storage. As thermal radiation for most Earth-bound heat transfer processes operates at "room temperature," the electromagnetic wavelength band of interest is between 3 and 40 µm (Mid-Infrared and Terahertz-wave). This micron-scale characteristic light regime motivates formulating technologies interacting with micron-scale and sub-micron patterned engineered materials – metamaterials.

The various approaches in photonic metamaterials have included multilayer heterostructures, gratings, nanoparticles, nanotubes, split-ring resonators, and many other nano/micro-scale patterned elements. To achieve the physical manifestation and practical utilization of these nano/microstructures, the various additive and subtractive micro-manufacturing processes have involved chemical vapor deposition (CVD), physical vapor

DOI: 10.1201/9781003409090-4

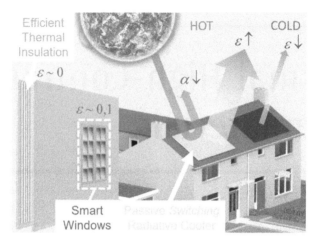

FIGURE 4.1 Technology processes implementing the next-generation high-contrast photonic metamaterials, such as more efficient thermal insulation, smart thermal windows, and passive emissivity switching radiative cooler coatings. Symbols α is solar absorptivity, and ε is infrared emissivity.

deposition (PVD), photolithography, nano/micro-milling, and many other "top-down" and "bottom-up" processes with atomic precision. These metamaterials are rooted in the materials chemistry from basic elements: Crystals, oxides, organic polymers, 2D materials, and novel topological or spin transport materials. Our coverage of materials' optical characteristics implemented in metamaterial form and function can inform the reader on best practices and modern achievements for net-zero or low-power thermal management.

In this review, we introduce the reader to the working principles of thermal radiative metamaterials, within the scope of nano/micro-materials technology design, development, characterization, and demonstration of capabilities/processes. The focus term on high-contrast emphasizes the near-perfect radiative performance and dual-phase tuning qualities of metamaterials. The designs are guided by two topical thrusts: High-Contrast Gratings (HCG) and variable thermal radiative property switching. Fundamental mechanisms of light-matter interactions, materials selection, and methods of metamaterial characterization are presented first. Theory of Fabry-Pérot interference in thin films and HCG, and their applications across the broad electromagnetic spectrum are discussed. We cover various materials, designs, and devices to achieve thermal emissivity switching. A brief overview of the future outlook in the field and impact toward a low-carbon society is summarized.

4.2 FUNDAMENTALS OF THERMAL METAMATERIALS

4.2.1 Principles of Thermal Radiation

An object at a uniform temperature (T) will emit electromagnetic waves or photons to its surroundings on every surface. The law that governs this emissive cooling by releasing the object's internal energy is the blackbody power law. The associated emissive power (E) is determined by its radiation from the object's surface, given by $E = \varepsilon\sigma T^4$. σ is the Stefan-Boltzmann constant. ε is the emissivity of the surface, where a perfect blackbody has $\varepsilon = 1$. On the other hand, a perfect mirror has $\varepsilon = 0$, which is typically attributed to polished metal surfaces. Figure 4.2 presents spectral emittance curves for obtaining idealized emissivity against an associated blackbody's emissive power spectrum near "room temperature." In ordinary applications, achieving near-perfect blackbody or near-perfect mirror has converged to within 2% in broadband visible and infrared range, i.e. $\varepsilon = 0.98$ and $\varepsilon = 0.02$, respectively.

Another aspect of near-perfect thermal-optical properties is its absorptivity (α), which corresponds to the blackbody temperature of irradiation. A common example of irradiation is that from the Sun, where the Sun's temperature of approximately $T = 5800$ K corresponds to a wavelength band between 0.3 and 3 µm (UV and Short-IR). Both thermal absorptivity and emissivity are spectrally-totaled across absorptance and emittance band wavelengths ($\lambda_1-\lambda_2$), and hemispherically across polar (θ) and azimuthal (ϕ) angles. Commonly, radiative emissivity is diffuse, with little variation along angle. In some considerations, thermal radiative properties

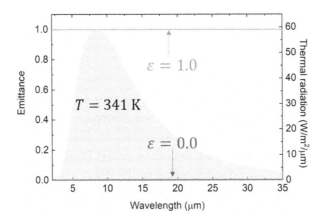

FIGURE 4.2 Emittance spectra for zero and perfect thermal emissivity against a "room temperature" blackbody emissive power spectrum.

are spectral, directional, or normal – refer to assumptions behind Kirchhoff's Law of Radiation (Z. M. Zhang 2020).

The constituent physics of thermal radiation is electromagnetic (EM) wave propagation. EM waves can also be evanescent, such that coupling can occur between dipole-dissipating bodies in near-field. Near-field thermal transport in the regime of sustained nanoscale vacuum gaps can yield an apparent emissivity beyond unity ($\varepsilon > 1$) (Z. M. Zhang 2020; Yang et al. 2023). However, thermal metamaterials and devices in near-field are not included in the scope of this chapter.

4.2.2 Electromagnetic Interference and Oscillations

The mechanisms behind thermal metamaterials are derived from near-field photon transport. Metamaterials of the first order – multilayers – are designed by coupling EM surface waves between material interfaces through continuum at short distances. For example, the coupling between two parallel and thin reflecting surfaces across a sub-wavelength cavity layer, called a Fabry-Pérot (FP) interferometer, results in near-perfect transmission or reflection (Z. M. Zhang 2020; L. P. Wang, Basu, and Zhang 2012). Figure 4.3(a) illustrates the basic three-layer principle of FP interferometry, and Figure 4.3(b) shows the calculated phase-shift transmittance for the first case of 90% reflecting surfaces about a cavity having a refractive index of $n = 2.0$.

The theory behind FP interference rests in the EM wave phase shift across the cavity layer. EM waves in the cavity that are in phase are constructively interfering, wherein the number of propagating round-trips is a whole number. Formally, the phase shift ψ is quantified as,

$$\psi = 2\pi dn\nu \cos\theta \tag{4.1}$$

where d is the thickness of the cavity, n is the cavity's complex refractive index, ν is the EM wave frequency in wavenumber, and θ is the light beam incident angle (Z. M. Zhang 2020). When $2dn\nu\cos\theta$ matches an integer (m), we call the cavity light as being in phase. At these mode resonances, the light is perfectly transmitted out of the cavity. This assumes that n is a real number – any non-zero absorption coefficient in the refractive index, κ, yields a decaying EM field, generally proportional to $e^{-\kappa d}$. The second example in Figure 4.3(b) shows a lossy cavity having non-zero κ, where the phase-matched transmittance is reduced. In the next section, we discuss

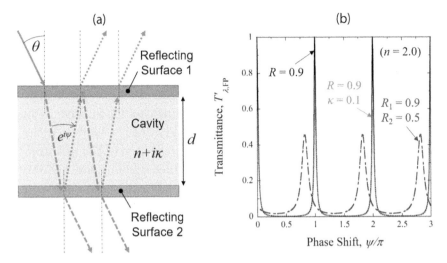

FIGURE 4.3 (a) Schematic of a Fabry-Pérot interferometer, with a cavity of complex refractive index of $n+i\kappa$ and thickness d. Light that enters the cavity at incident angle θ is multiply reflected and transmitted with phase shift $e^{i\psi}$. (b) Transmittance calculations for symmetric reflecting surfaces ($R = R_1 = R_2 = 0.9$, $\kappa = 0$), a lossy cavity ($R = 0.9$, $\kappa = 0.1$), and asymmetric reflecting surfaces ($R_1 = 0.9$, $R_2 = 0.5$, $\kappa = 0$).

cavity and reflecting surface materials selection and their optical properties that permit near-perfect interference at thermal wavelengths.

The quality and breadth of block-band frequencies of near-perfect reflection can be increased by stacking successive alternating layers of FP interferometers at half-integer phase shifts, called a Distributed Bragg Reflector (DBR) (Blankenship, Adams, and Zhang 2021; Z. M. Zhang 2020). The layer thicknesses d are dictated by the quarter-wave equation at normal incidence,

$$d = \frac{\lambda_0}{4n} \tag{4.2}$$

where λ_0 is the center wavelength for maximum reflectance. A DBR with a high number of alternating layers (N) can achieve ultrahigh reflectance exceeding 99.9% ($N = 40$). An application of a DBR is in wavelength laser diodes, where its composition is alternating-doped and epitaxied III-V semiconductors, i.e. GaAs/AlGaAs (Chang-Hasnain 2000). As the theory behind optical resonance in band-transparent dielectrics or non-conductors is simple, the mechanism is slightly more complex between dielectrics and metals.

At the interface where one side is electron-rich (i.e., metals), plasmonics grows from localized light-matter interactions at the interface where one side is electron-rich (Maier 2007). Plasma oscillation packets, called plasmon polaritons, can be excited when a photon matches the frequency and momentum of the electron gas at their plasma frequency. Ways to excite plasmon polaritons include Attenuated Total Reflection (ATR) prisms or waveguides, tunnel junctions, and nanopatterned metamaterials (Z. M. Zhang 2020). Synthetic means to excite and tailor plasmon polaritons include metallic and semiconductor nanoparticles, metal-dielectric multilayers, plasmonic gratings, and many other resonant metal nano/microstructures (Zheludev and Kivshar 2012). Under plasmonic resonance, rather than ultrahigh EM wave transmission or reflection, the photon is fully absorbed by the plasma. Via Kirchhoff's Law, plasmonic resonance also applies to photon emission. With the right design and materials, plasmonic emission metamaterials can serve as a dual-purpose optical interferometer, thereby achieving high-contrast radiative properties.

4.2.3 Metamaterials in Applications

Thermal metamaterials are implemented in many practical modern applications and devices: Anti-reflection coatings, laser diode gain and thermal management, on-chip photonic integrated circuits, sensors for biomolecules, telecommunication fibers and mirrors, and many more (Y. Li et al. 2021; J. C. Kim et al. 2021). The technology improvements in the last few decades have allowed for more efficient, scalable, sustainable, and lower cost integrated metamaterials. A main impact of realized metamaterials is miniaturization of electronic and mechanical devices due to their nano/micro-scale componentry. Secondary implications of metamaterials are their augmented energy efficiency across photosensitive electronics, such as solar cells, thermophotovoltaics, and electromechanical energy harvesters (Shin et al. 2018). The most important aspect involving photonic metamaterials is their ability to enable energy transport in unconventional ways: Negative index of refraction that was originally implied by John Pendry, as well as object cloaking or apparent invisibility, one-way transmission sorted by wavelength or beam momentum, passive thermodynamic radiative cooling under the Sun to the surroundings, and some nonlinear optical processes (Pendry 2000; Cai et al. 2007; Smith, Pendry, and Wiltshire 2004; Fan and Li 2022; Almeida, Bitton, and Prior 2016). We must prioritize efforts to develop well-engineered metamaterials that implement next generation materials to meet global demands for structural and personnel energy efficiency, sustainable power harvesting and storage, medical sensing and diagnosis, and the vast array of electronic nano/micro-devices.

4.3 MATERIAL AND THERMAL CHARACTERISTICS

4.3.1 Thermal Materials

Thermal properties of metamaterials primarily lie in their constituents' electronic band structures. Most simply, dielectric insulators have a wide bandgap between valence and conduction bands for Fermi energies corresponding to thermal wavelengths, which range between a fraction of electron volts (~0.1 eV) to a few electron volts (~1 eV). Elemental examples of thermal dielectrics are Silicon (Si), Germanium (Ge), and Diamond (C) (Harris 1999). Hallmarks of good thermal dielectrics are high-order crystallinity, high phonon conductance, high electron mobility, and high electric field permittivity relative to vacuum (Z. M. Zhang 2020). Some III-V semiconductors with Zinc Blende lattice structure also exhibit excellent dielectricity, such as Gallium Arsenide (GaAs) and Zinc Selenide (ZnSe) (Harris 1999). Not all semiconductors are dielectric in the thermal wavelengths of interest.

Semiconductors in the oxide family, crystalline silicon dioxide (SiO_2) and titanium dioxide (TiO_2) as the most common examples, are poor dielectrics for thermal radiation. While SiO_2 (quartz glass) is transparent in the visible wavelengths due to its very wide bandgap (~9 eV), it blocks infrared light due to its lattice oscillations captured in photon-phonon absorption (Kitamura, Pilon, and Jonasz 2007). This three-energy pseudoparticle interaction among photons, phonons, and electrons can generally be pictured as series of indirect transitions of the conduction band. Photon absorption across an electron bandgap can be described by the Lorentz model, a harmonic oscillator spectral function analogous to the classical mass-spring-damper dynamics. Metals on the other hand are simply represented as high-density plasma entities, where the Drude model spectral function represents a classical mass-damper system for photon-electron absorption in the conduction band (Z. M. Zhang 2020).

Interestingly, certain classes of optical materials can exhibit both dielectric and metallic states. These solid-phase and state-reversible materials are called metal-to-insulating phase transition oxides (PTO). Recent advances in a near-room temperature PTO Vanadium Dioxide (VO_2) have excited applications in temperature-sensing and switchable infrared radiation management, optical communication modulators, and neuromorphic phase memory computing (Liu et al. 2018; Shi et al. 2019). We elaborate the key features of VO_2 and its integration in thermal metamaterials in Section 4.5: High-Contrast Thermal Switching. Other kinds of PTO in thermal metamaterials that switch with temperature and/or voltage are also elaborated.

4.3.2 Thermal Metrology

In the quest to better understand optical parameters of materials, various spectroscopic metrology methods have been developed and utilized over the past few decades. For relevancy to thermal metamaterials characterization, the methods are applied to solid and monolithic surfaces in the visible to infrared wavelengths. An efficient method of obtaining spectral radiative properties is Fourier-transform spectroscopy, particularly that in the infrared range (FTIR). An FTIR passes a beam containing multiple frequencies through or across a sample to an infrared intensity detector. Frequency-dependent coherent absorption in the detector is achieved by Fourier transform algorithms working in sync with a moving Michelson interferometer. The FTIR is exceptional in resolving signals outside incoherent thermal noise, but its sensitivity to atmospheric gas vibrational absorption may be picked up in un-evacuated chambers. Customizable FTIR with capabilities for sample thermal and electronic control are a critical tool for developers of thermal metamaterials (C. Chen et al. 2022; Yang et al. 2021). Figure 4.4(a, b) illustrates an FTIR pathway for measuring thermally sensitive coatings and thin films, especially FP interferometers that generate low signal-to-noise (high thermal absorption) (L. P. Wang, Basu, and Zhang 2012).

Equally valuable spectroscopic techniques involve monochromatic emission and detection of polarized light. Despite lower rate of data acquisition, monochromatic spectrometry using a diffraction grating is most frequently employed in integrating spheres for measuring radiative properties of diffuse surfaces. This technique is limited to EM frequencies outside the thermal noise regime, as optical components at room temperature insert incoherent signals. For specular radiative properties, spectroscopic ellipsometry employs a monochromatic polarized light source and a three-axis analyzer-detector to obtain reflectance off surfaces (Z. M. Zhang 2020).

Metamaterial characterization methods *in situ* are less spectroscopic: Thermal performance is either assessed by integrated radiative properties or measured by far-field calorimetry. A solar reflectometer or infrared emissometer is used to obtain the total hemispherical properties of a

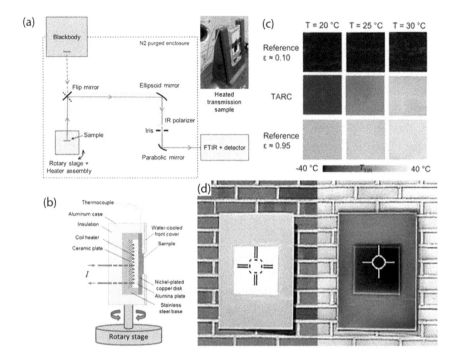

FIGURE 4.4   (a) Thermal emissometry technique using a FTIR spectrometer and blackbody calibration for (b) a heated Fabry-Pérot thin film sample. (Adapted with permission from L. P. Wang, Basu, and Zhang 2012.) Far-field thermal calorimetry of (c) a thermochromic variable emissivity sample (TARC). (Adapted with permission from Tang et al. 2021.) (d) a large-scale passive radiative cooler paint (square) against its reference (rectangle) under visible (left) and infrared (right) images. (Courtesy of Purdue University/Joseph Peoples.)

surface viewed from an integrating sphere aperture (Pettit 1978). This approach is limited to mostly diffuse radiating surfaces, whereby thermal performance of highly specular surfaces such as gratings may not be well represented. Far-field calorimetry is a simple object imaging approach using an infrared camera with high resolution detectability in the far-infrared (LWIR) wavelengths (Tang et al. 2021; X. Li et al. 2020). Far-field calorimetry can resolve spatial and temporal variations in thermal samples, as shown in Figure 4.4(c). They can also compare apparent emissivity of large-scale coatings against its absorptivity using a research camera, shown in Figure 4.4(d). Once the temperature and environmental conditions around samples are well-controlled, these non-destructive thermal characterization methods are highly valuable in evaluating metamaterials in practical thermal applications.

4.4 THERMAL HIGH-CONTRAST GRATINGS

4.4.1 Origin of High-Contrast Gratings

High-Contrast Gratings originated upon the question: What if Fabry-Pérot interferometry can be dispersed across 3D spatial order? Instead of the axisymmetric symmetry about the optical or normal-to-the-surface axis of DBR, the high-low refractive index repetition is lateral. In the mid-2000s, Professor Connie Chang-Hasnain devised the High-Contrast Grating (HCG) to achieve ultrahigh reflectance on an actuated micro-electromechanical (MEMS) device (Huang, Zhou, and Chang-Hasnain 2007). These thin suspended HCG allowed emission and wavelength tuning of semiconductor vertical cavity surface emitting lasers (VCSEL), shown schematically in Figure 4.5(a, b) (Chase et al. 2010). In NIR/SWIR VCSEL applications,

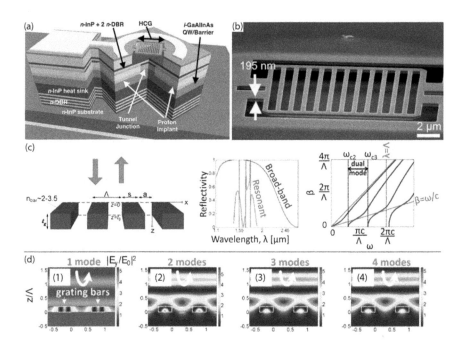

FIGURE 4.5 (a) Schematic of High-Contrast Grating (HCG) in a vertical cavity surface emitting lasers (VCSEL) diode, with (b) fabricated suspended HCG device. (c) Design principles of HCG showing its refractive index and dimensional parameters, broadband reflectivity, and dual-mode region and higher (crossings beyond ω_{c3}) for HCG dispersion. (d) The multi-modal Fabry-Pérot round-trip showing transverse-electric fields constructively interfering above the HCG plane. (Adapted with permission from: Chase et al. 2010; Karagodsky and Chang-Hasnain 2012; Karagodsky, Sedgwick, and Chang-Hasnain 2010.)

HCG have been made of wide direct bandgap III-V semiconductors, such as GaAs, InP, AlGaAs, and others (Qiao, Yang, and Chang-Hasnain 2018). In more recent advancements, Silicon-based HCG have been made for long-wavelength (LWIR) telecom lasers. The unifying characteristic of HCG materials is their high refractive index n, with low absorption coefficient κ. The high refractive index n of the HCG broadens its ultra-reflectance band, shown in Figure 4.5(c) (Karagodsky and Chang-Hasnain 2012). The HCG's performance is reduced with absorption coefficient κ, which can be illustrated by the second Fabry-Pérot example in Figure 4.3(b). The general rule of thumb for an effective HCG is: $n > 3$ and $\kappa < 10^{-2}$.

The physics of HCG was succinctly explained by Karagodsky and Chang-Hasnain: In principle, HCG are at-wavelength gratings that enable multi-mode destructive interference. Through Rigorous Coupled Wave Analysis (RCWA) calculations, the HCG layer was found to mismatch the phase of incoming and outgoing propagating waves, called a Fabry-Pérot round trip (Karagodsky and Chang-Hasnain 2012). These round trips manifest as an interference plane at double grating height, for at least dual modes, shown in Figure 4.5(d) (Karagodsky, Sedgwick, and Chang-Hasnain 2010). The bands' waveguide array modes are obtained from the following conjunctive dispersion equation,

$$k_g \tan\left(\frac{k_g f \Lambda}{2}\right) + n^2 k_a \tan\left[\frac{k_a (1-f)\Lambda}{2}\right] = 0 \qquad (4.3a)$$

$$\beta^2 = k_0^2 - k_a^2 = k_0^2 - n^2 k_g^2 \qquad (4.3b)$$

where k_g and k_a are the grating and air gap wavevectors for each mode (Karagodsky and Chang-Hasnain 2012). In this equation, the grating period (Λ) and the solid-to-period ratio or duty cycle (f) are the key qualifying dimensional attributes, along with the grating refractive index (n). Solving this equation would suggest that the HCG grating period is about 2–3 times smaller than the center wavelength of its reflected light. Both the grating height and duty cycle are consistently close to half the size of the grating period. Both unidirectional or 1D HCG, and 2D HCG in rectangular or circular pattern can achieve reflectance exceeding 99% (Qiao et al. 2015).

Recent findings by Marciniak et al. have expanded HCG designs to other shape factors, such as 1D trapezoids, ellipsoidal-tip fins made of polymers, and even simulated cross-sections of heads of state – verifying the primary importance of the lateral electric field waveguide effect with the alternating high and unity refractive index components (Marciniak et al. 2016; Jandura et al. 2020; Marciniak et al. 2020). While HCG can provide a thin and simple reflector, discovery and fabrication of more complex HCG patterns are needed to account for both transverse magnetic (TM) and transverse electric (TE) polarizations and to achieve angular independence across broad wavelengths.

4.4.2 Infrared High-Contrast Gratings

Several recent works in the last decade have applied the theory of HCG to infrared wavelengths corresponding to thermal radiation. In Figure 4.6(a), Hogan et al. surveyed several grating and substrate materials suitable for LWIR/Far-IR transparency (Hogan et al. 2016). GaAs, Silicon, and Germanium are good candidates for fabricating HCG, as their refractive indices are 3.3, 3.5, and 4.0, respectively. Only GaAs has a high absorption coefficient ($\kappa = 0.8 \times 10^{-3}$) at the 10 μm center wavelength. Therefore, only Si and Ge are considered good candidates for IR HCG. The micro-fabrication process of hard crystals such as Si and Ge requires deep reactive ion etching of the negative space developed by UV photolithography, as shown in Figure 4.6(b); high quality Ge HCG are produced, as shown in Figure 4.6(c). When calculated and measured for their reflectance, a monolithic layer of Ge HCG produced a reflectance band between 10 and 12 μm – not broad enough for "room temperature" thermal radiative insulation.

Since the Planck blackbody emission function has a tail toward longer wavelengths, radiative properties must extend farther into the Far-IR spectrum for more effective thermal emission control. Our work, shown in Figure 4.6(d), combined a monolithic HCG with a FP cavity (R. Z. Zhang and Araki 2023). Using computational optimization and sensitivity treatments, we have summarized the best-case parameters for this multilayer thermal coating: The HCG is Ge, and the FP cavity is IR-transparent cubic salt KBr ($n = 1.55$). The HCG grating period is 2.5 times smaller than the center wavelength (10 μm). While the filling ratio of 0.6 and the HCG thickness of half-period are not out of the ordinary, the combination involved significant dimensional changes in the FP cavity. The cavity thickness is approximately 50% more than calculated by the quarter-wave equation given in Equation (4.2). The result of our parameter optimization is a very low normal emissivity of $\varepsilon = 0.0085$, a magnitude lower than the polished and pristine aluminum surface of $\varepsilon = 0.03$ (Figure 4.6(e)).

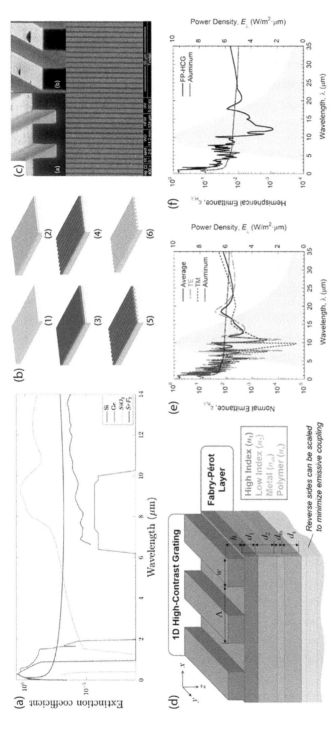

FIGURE 4.6 (a) Transparent infrared (LWIR/Far-IR) materials absorption/extinction coefficient (κ) spectra for high-index Si and Ge, and a low-index fluoride (SrF₂). SiO₂ is an infrared absorber. (b) Pattern process of HCG, showing (1) low-index substrate, (2) high-index coating, (3) photomasking, (4) UV pattern exposure and wash-off, (5) deep-reactive ion etching, and (6) final HCG coating, where (c) a high-quality HCG sample can be made. (Adapted with permission from Hogan et al. 2016.) (d) A combined Fabry-Pérot and HCG coating can be made on a flexible substrate with a backside that further decouples radiative absorption. (e) The normal emittance spectra for TE/TM polarizations, their average, compared to pure aluminum, and (f) its hemispherical emittance spectrum. (Zhang and Araki 2023, wih permission.)

This HCG-FP combined design was also able to achieve a hemispherical total emissivity of $\varepsilon^\frown = 0.014$.

This ultralow emittance design in broad LWIR/Far-IR can signal an effective thermal radiative insulator provided a vacuum is preserved between two far-field surfaces facing each other. Moreover, the HCG-FP can be designed asymmetrically, where the two sides' maximum reflectance bands do not overlap in the blackbody emission spectrum. The HCG can also be in a 2D rectangular pattern, but its performance is not much better than 1D HCG. The prime advantage of 1D HCG is their polarization contrast, where coupling between perpendicularly oriented gratings can further minimize coherent thermal radiative coupling (Greffet and Henkel 2007).

4.4.3 Outlook on High-Contrast Gratings

Many photonic metastructures can also achieve ultrahigh reflectance, or near-perfect emittance. Examples include metal-insulator-metal micropatterned surfaces, 2D material heterostructures, and polaritonic resonators (H.-T. Chen, Taylor, and Yu 2016). An analogous approach to the HCG is the Zero-Contrast Grating (ZCG). Devised by Prof. Robert Magnusson, the ZCG is based on waveguide-mode resonance or leaky Bloch mode propagation (Magnusson 2014). The overall scheme that the ZCG captures is better suited for ultrahigh broadband transmittance, where LWIR polarization filters have been demonstrated in a Ge ZCG on a ZnSe antireflecting substrate (Ko et al. 2023). Either HCG or ZCG can provide a useful broadband and high-performance reflection or transmission metasurface, and the realization in thermal radiative control systems is yet to be seen.

4.5 HIGH-CONTRAST THERMAL SWITCHING

4.5.1 Passive Thermal Emission Switching

Various metamaterial structures and physical mechanisms have been introduced which induce thermal radiative switching, especially utilizing VO_2 with its reversible metal-to-insulating phase transition temperature at 341 K. These VO_2 structures include multilayers and periodic gratings to enhance the contrast in emissivity, absorptivity, and heat flux as shown in Figure 4.7.

The FP resonance arrangement can generate at-wavelength near-perfect to zero emissivity while the dielectric cavity has thermally driven metallic boundaries (Taylor, Yang, and Wang 2017; Ghanekar, Xiao, and Zheng 2017).

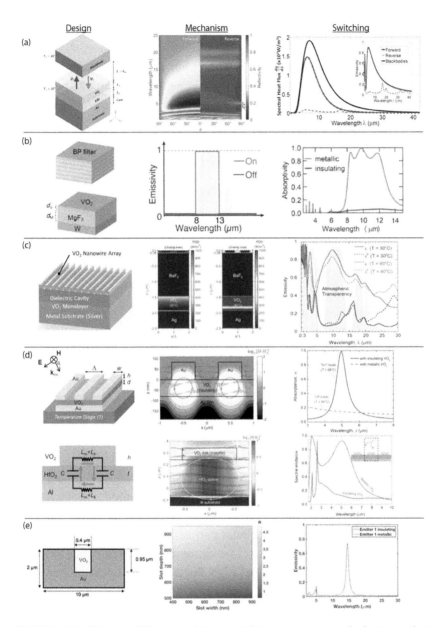

FIGURE 4.7 Diverse VO$_2$ emissivity switching metamaterial designs, their electromagnetic wave absorption mechanisms, and spectral evidence of switching: (a) A two-body Fabry-Pérot VO$_2$ thermal rectifier, (b) a Fabry-Pérot VO$_2$ passive radiative cooler with a photonic crystal solar blocker, (c) an optimized VO$_2$ nanowire grating with a sub-monolayer VO$_2$, (d) metal-insulator-metal or inverse plasmonic resonance VO$_2$ switching emitters, and (e) a VO$_2$-in-groove surface plasmon-based heat rectification emitter. (Adapted with permission from: Ghanekar, Xiao, and Zheng 2017; Ono et al. 2018; Araki and Zhang 2022a; H. Wang, Yang, and Wang 2014; Long, Taylor, and Wang 2020; Audhkhasi and Povinelli 2019.)

In the case by Ghanekar, Xiao, and Zheng, shown in Figure 4.7(a), one can achieve a large thermal rectification, defined as the ratio of difference between the forward and reverse heat flux between blackbodies to their reverse heat flux. Since it develops the temperature difference between those two plates, one being the VO_2 FP structure, the reverse heat flux from the cold blackbody is suppressed while VO_2 is in its insulating phase (Ghanekar, Xiao, and Zheng 2017). This concept of a two-body heat flux exchange was also proposed by Ono et al. for a passive and switching radiative cooling device where cooling power is ON during the hot ambient temperature, as shown in Figure 4.7(b). Their proposed structure is composed of a multilayered solar reflecting-IR transparent filter suspended above the VO_2 thermal emitter (Ono et al. 2018). In this two-body example, it can be challenging to separate the filter and the emitter where heat flux exchange at thermal equilibrium is minimized. One possible solution is to deposit a high index Si slab on top of the VO_2 emitter (H. Kim et al. 2019).

Numerous developments have used diverse photonic metastructures, including nanowire gratings, nanoparticles, and biomimic structures for passive radiative cooling, which emits heat through atmospheric window of 8–13 µm while rejecting the solar irradiance with high reflectivity (Lee et al. 2023). This influenced our horizontally-aligned VO_2 nanowire design shown in Figure 4.7(c), where the effective medium temperature-switching reflection surface covers a low-index BaF_2 FP cavity (Araki and Zhang 2022a). This study also unveiled how a solid VO_2 layer below the low-index FP cavity leads to a higher total emissivity contrast. The mechanism can be explained by Figure 4.3(b), where the FP phase shift modes are destroyed by asymmetric reflecting surfaces – the VO_2 "sub-monolayer" forces refractive index matching and provides an additional source of radiative attenuation. Despite its excellent performance in thermal radiative switching, its solar absorptance remains high due to the constant absorption coefficient of VO_2 in the Visible wavelengths.

As opposed to FP-based multilayers, sandwiches, and grating structures, plasmonic resonance can induce the infrared switching because of VO_2's Lorentz to Drude behavior at cold to hot temperatures. As shown in Figure 4.7(d), the insulating phase of VO_2 is transformed as the dielectric cavity surrounded by the metallic grating and sub-metallic plate to induce magnetic resonance such that near-unity absorptance is fulfilled at the insulating phase (H. Wang, Yang, and Wang 2014). In a similar approach, the metallic VO_2 is employed as the top metallic boundary of the dielectric cavity for magnetic polariton (MP) resonance generation, whose peak can be explained by the frequency-dependent LC circuit model near-field conducting boundaries (Long, Taylor, and Wang 2020). The MP peak emittance can be explained by the frequency-dependent LC circuit model about near-field conducting boundaries (Z. M. Zhang 2020). The thicker VO_2 as a dielectric cavity in the former case and thicker dielectric cavity in the latter case is required to utilize these magnetic resonances in thermal radiative wavelength at around 8 to 10 μm. Similar plasmonic gratings were presented with a 2D VO_2 grating with dielectric spacer but with lower emissivity turn-down (Ito et al. 2018; Sun et al. 2022).

The MP resonance can also be induced within the air or vacuum grating groove with a periodic VO_2 deep grating at its metallic phase (Audhkhasi and Povinelli 2019; Guo et al. 2020). In the approach by Audhkhasi and Povinelli, shown in Figure 4.7(e), the surface plasmon polariton (SPP) resonance is employed for narrowband emittance switching using VO_2 which is embedded in a gold periodic grating structure. They report a high rectification factor of 20 from a standalone radiative structure in far-field. The high rectification ratio was determined by optimizing the VO_2-filled grating groove height and width such that its emissivity due to SPP excitation is close to unity at its insulating state but near-zero at its metallic state (Audhkhasi and Povinelli 2019).

Passive temperature-regulating emissivity switching was demonstrated in various designs of a small component of VO_2 surrounded by either a dielectric spacer layer by FP resonance or a micro-patterned metal metastructure by plasmon-polaritonic resonance. In the overall consideration, ease of fabrication of these structures is most important, as reducing multistep photolithographic patterning, the number of deposition-thermal anneal cycles, and preventing material cross-contamination are all important in producing high-quality VO_2-containing metamaterials. The enabling of PTO or VO_2 allotropes such as nanowires, nanoparticles, or even atomically thin 2D sheets can help with novel metamaterial fabrication and intrinsic optical properties.

4.5.2 Electrochromic Switching

Optical property switching can also be accomplished with electrical voltage gating across classes of PTO. The reversible mechanism is due to action potential-driven lattice change rather than Joule thermal heating of the slab. For example, Tungsten Trioxide (WO_3) can shift colors in the visible wavelengths on input of ~1 VDC gating with Li^+ cation insertion

(X. Zhang et al. 2019). This color change can be impactful for smart self-shading windows. However, adjusting "color" for infrared wavelengths from 3 to 40 μm is much more impactful for radiation exchanges to objects' natural surroundings. The quest to find an electrochromic material in the Mid- to Far-IR has been challenging. X. Zhang et al. (Figure 4.8(a)) have demonstrated a positive and negative voltage-biased WO_3/$LiTaO_3$/NiO electrochromic switching multilayer with decent emissivity

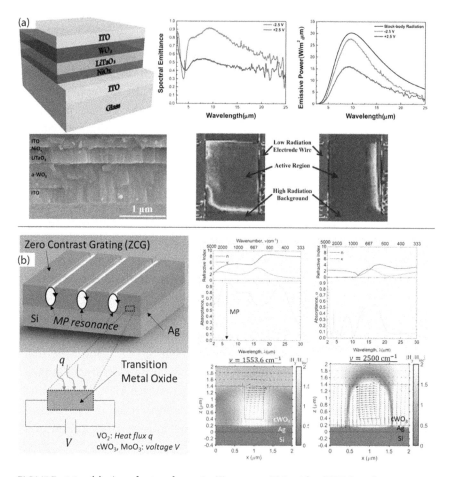

FIGURE 4.8 (a) An electrochromic Tungsten Trioxide (WO_3) voltage-gating variable emissivity metamaterial, showing positive and negative voltage bias thermal emission tuning, its heterostructure micrograph, and thermal calorimetry images. (Adapted with permission from X. Zhang et al. 2019.) (b) A WO_3 polariton-switching Zero-Contrast Grating (ZCG) variable emissivity metamaterial, showing magnetic plasmon-polariton (MP) high-emittance mode when "on" and ZCG low-emittance mode when "off." (Araki and Zhang 2023, with permission.)

contrast of $\Delta\varepsilon = 0.37$ in the LWIR/Far-IR wavelength band 8 to 14 μm. The low voltage infrared switching is promising, but more can be done to improve electrochromic phase transition oxide crystallinity and optical design to maximize interference at both states.

Returning to the topic of the Zero-Contrast Grating (ZCG) discussed in Section 4.4.3, we utilized this design to achieve radiative emission switching through the activation of MP resonance (Araki and Zhang 2023). In the insulating phase, where no voltage is applied, the crystalline WO_3 ZCG is in the leaky mode, where thermal emittance is minimized. With a low applied voltage (~1.0 VDC), the metallic phase of cWO_3 excites resonance of MP, which are moded narrowband emittances described previously in Section 4.5.1. Under this Bloch-polaritonic switching mechanism, not related to FP interference, the emissivity contrast can reach to $\Delta\varepsilon = 0.75$, higher than an optimized VO_2 FP structure (Araki and Zhang 2022a). We also demonstrated higher emissivity contrast in a VO_2 ZCG ($\Delta\varepsilon = 0.89$), and also in an up-and-coming electrochromic Molybdenum Trioxide (MoO_3) ZCG ($\Delta\varepsilon = 0.97$). More work to develop a metamaterial that switches between HCG/ZCG and MP is encouraged, as the grating groove design in a phase transition monolayer is simple and scalable.

4.5.3 Combined High-Contrast and Switching Metamaterials

A natural question that comes with the topical combination of HCG and PTO is: Can a PTO be a HCG? It is possible with a HCG designed with an ultra-reflectance bandwidth that is concurrent with the transparency window of insulating phase PTO. In VO_2, the transparency window is neither wide enough to cover the broad Planck blackbody emission distribution nor low enough in its absorption coefficient. Therefore, a multi-component heterostructure is needed to combine HCG capability with infrared emissivity switching. The design involves a topmost layer consisting of a broadband HCG reflector in the Visible and Near-IR wavelengths. The underlying layers consist of more-or-less the same switching FP multilayer with a PTO gate-absorber pair surrounding a low refractive index cavity. The HCG must remain as an IR-transparent Silicon or Germanium topping layer, with the latter material further constrained by its Visible wavelength opacity and temperature-dependent opacity above 100°C (Li 1980; Zhang et al. 2020). In our study, shown in Figure 4.9(a), a Silicon 1D HCG above a VO_2/BaF_2 heterostructure provides superb normal reflectance in the Visible/Near-IR or atmospheric sunlight regime, while maintaining the IR switching function of the under-layers (Araki and Zhang 2022b).

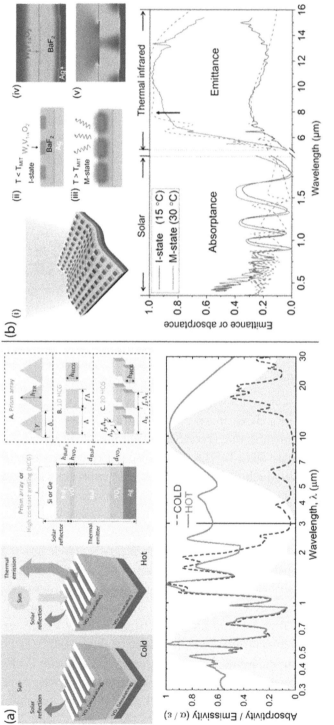

FIGURE 4.9   (a) A stacked Si 1D HCG-on-variable emissivity VO$_2$ multilayer with reduced solar absorptivity. The top layer can also be a diffractive micro-prism array. (Araki and Zhang 2022b.) (b) A Tungsten-doped VO$_2$ metasurface on a BaF$_2$ flexible substrate, which also shows switchable passive radiative cooling. (Adapted with permission from Tang et al. 2021.)

A key contributor to the performance is a secondary propagating layer of BaF_2 or a similar low-index layer between the switching FP and HCG array. The moderate solar absorptance of $\alpha = 0.57$ simultaneous to the wide emissivity contrast of $\Delta\varepsilon = 0.71$ can potentially be used as a tunable passive radiative cooler.

Tang et al. have demonstrated a similar tunable passive radiative cooler with contrast of $\Delta\varepsilon = 0.70$, while lowering the solar absorptance to $\alpha = 0.25$ (Figure 4.9(b)) (Tang et al. 2021). Their approach to use Tungsten-doped VO_2 not only enabled a lower metal-to-insulating transition temperature near 25°C but may have also decreased the absorption cross-section by implementing 2D HCG-like micro-islands at the optimal filling ratio. While the exposed BaF_2 low-index layer may present viability challenges in the real environment, the thin structure's encapsulation within IR-transparent and flexible polymers may hold promise to a scalable and sustainable tunable passive radiative cooler.

4.6 DISCUSSION OF METAMATERIAL APPLICATIONS AND FUTURE OUTLOOK

From this review, we realize several technology development trajectories are needed to realize near-perfect optical properties and switching of thermal properties. The fundamental exploration of materials chemistry is needed to identify relationships in crystallography, deposition, synthesis, and thermal considerations to their optical properties across the EM spectrum. The discovery of a low-refractive index and broadband infrared transparent solid at room temperature is commendable. Unearthing a reversible phase transition solid that can switch between pure dielectric and electron-rich metal near room temperature will be exceptional. Tuning the bandgap or controlling the lattice structure can pinpoint a desired phase transition temperature or voltage. Further capabilities relating to condensed matter physics of energy pseudoparticles, such as electron spin and coupling, topological phases, and room temperature superconductivity, may also emerge as critical contributors to future metamaterials.

The next step is to improve the building blocks of metamaterials. We presented innovations in applying the unifying optical mechanism of interferometry to photonic metamaterials with careful selection of component chemistry. Materials identifying as broad-wavelength infrared transparent and understanding their optical phase shift mechanisms are critical to metamaterial constituents. We also note that the expression "contrast" terminologically intersects with "phase" across various contexts, which may motivate other phase-related optical physics studies. Understanding the behavior of other temperature or electrical phase transition materials are a priority, as factors such as deposition crystallinity, doping, and thermal processing need to be linked to a near-ideal reversible metal-to-insulating transition. Novel and unique methods in nano/micro-fabrication are needed to advance both the attainment and quality of complex meta-materials. This review's topical emphasis on 3D nano/micro-gratings motivates the development of highly controlled thin film deposition and patterning processes.

4.7 CONCLUSIONS

In this chapter, we assess the approaches in novel thermal metamaterials for their abilities to effectively control heat in a compact yet monolithic treatment, and to tailor radiative properties that adapt to changing environments. The key impact areas of these metamaterial technologies are in passive thermodynamic cooling surfaces, more efficient thermal insulation, and thermal switches. These engineered surfaces can also be used for non-thermal applications, such as infrared imaging and telecommunication, non-destructive evaluation, and high-power lasers. The global implications around these advancements are in lowered dependency on pumped-loop engines, improved waste heat recovery systems, safety and resiliency of heat generating electronics, and the general improvement of function in nano/micro-devices.

ACKNOWLEDGEMENTS

This work was supported by the Army Research Office and was accomplished under Grant No. W911NF-23-1-0165. The views and conclusions contained in this document are those of the authors and should not be interpreted as representing the official policies, either expressed or implied, of the Air Force Office of Scientific Research or the U.S. Government.

REFERENCES

Almeida, Euclides, Ora Bitton, and Yehiam Prior. 2016. 'Nonlinear Metamaterials for Holography'. *Nature Communications* 7 (1): 12533.

Araki, Ken, and Richard Z. Zhang. 2022a. 'An Optimized Self-Adaptive Thermal Radiation Turn-down Coating with Vanadium Dioxide Nanowire Array'. *International Journal of Heat and Mass Transfer* 191: 122835.

Araki, Ken, and Richard Z. Zhang. 2022b. 'Simultaneous Solar Rejection and Infrared Emission Switching Using an Integrated Dielectrics-on-VO$_2$ meta-surface'. *AIP Advances* 12 (5): 1–11.

Araki, Ken, and Richard Z. Zhang. 2023. 'Infrared Radiative Switching with Thermally and Electrically Tunable Transition Metal Oxides-Based Plasmonic Grating'. *Scientific Reports* 13 (1): 3702.

Audhkhasi, Romil, and Michelle L. Povinelli. 2019. 'Design of Far-Field Thermal Rectifiers Using Gold-Vanadium Dioxide Micro-Gratings'. *Journal of Applied Physics* 126 (6): 063106.

Blankenship, Morgan A., Kelsa D. Adams, and Richard Z. Zhang. 2021. 'Gradient-Index Metasurface Multilayer for Quasioptical Coupling of Infrared Detectors'. *Optical Engineering* 60 (10): 1–13.

Cai, Wenshan, Uday K. Chettiar, Alexander V. Kildishev, and Vladimir M. Shalaev. 2007. 'Optical Cloaking with Metamaterials'. *Nature Photonics* 1 (4): 224–27.

Chang-Hasnain, C. J. 2000. 'Tunable VCSEL'. *IEEE Journal of Selected Topics in Quantum Electronics* 6 (6): 978–87.

Chase, Christopher, Yi Rao, Werner Hofmann, and Connie J. Chang-Hasnain. 2010. '1550 Nm High Contrast Grating VCSEL'. *Optics Express* 18 (15): 15461–66.

Chen, Chuyang, Shin Young Jeong, Devesh Ranjan, Peter G. Loutzenhiser, and Zhuomin M. Zhang. 2022. 'Spectral Radiative Properties of Solid Particles for Concentrated Solar Power Applications'. In *Annual Review of Heat Transfer*, edited by Vish Prasad, Yogesh Jaluria, and Zhuomin M Zhang, 175–221. Begell House.

Chen, Hou-Tong, Antoinette J. Taylor, and Nanfang Yu. 2016. 'A Review of Metasurfaces: Physics and Applications'. *Reports on Progress in Physics* 79 (7): 76401.

Fan, Shanhui, and Wei Li. 2022. 'Photonics and Thermodynamics Concepts in Radiative Cooling'. *Nature Photonics* 16 (3): 182–90.

Ghanekar, Alok, Gang Xiao, and Yi Zheng. 2017. 'High Contrast Far-Field Radiative Thermal Diode'. *Scientific Reports* 7 (1): 1–7.

Greffet, Jean-Jacques, and Carsten Henkel. 2007. 'Coherent Thermal Radiation'. *Contemporary Physics* 48 (4): 183–94.

Guo, Yanming, Bo Xiong, Yong Shuai, and Junming Zhao. 2020. 'Thermal Driven Wavelength-Selective Optical Switch Based on Magnetic Polaritons Coupling'. *Journal of Quantitative Spectroscopy and Radiative Transfer*. 255 (11): 107230.

Harris, Daniel C. 1999. *Materials for Infrared Windows and Domes: Properties and Performance*. Vol. 158. SPIE Press.

Hogan, Brian, Stephen P. Hegarty, Liam Lewis, Javier Romero-Vivas, Tomasz J. Ochalski, and Guillaume Huyet. 2016. 'Realization of High-Contrast Gratings Operating at 10 μm'. *Optics Letters* 41 (21): 5130.

Huang, Michael C. Y., Ye Zhou, and Connie J. Chang-Hasnain. 2007. 'A Surface-Emitting Laser Incorporating a High-Index-Contrast Subwavelength Grating'. *Nature Photonics* 1 (2): 119–22.

Ito, Kota, Toshio Watari, Kazutaka Nishikawa, Hiroshi Yoshimoto, and Hideo Iizuka. 2018. 'Inverting the Thermal Radiative Contrast of Vanadium Dioxide by Metasurfaces Based on Localized Gap-Plasmons'. *APL Photonics* 3 (8): 086101.

Jandura, D., T. Czyszanowski, D. Pudis, M. Marciniak, M. Goraus, and P. Urbancova. 2020. 'Polymer-Based MHCG as Selective Mirror'. *Applied Surface Science* 527 (2): 146827.

Karagodsky, Vadim, and Connie J. Chang-Hasnain. 2012. 'Physics of Near-Wavelength High Contrast Gratings'. *Optics Express* 20 (10): 10888.

Karagodsky, Vadim, Forrest G. Sedgwick, and Connie J. Chang-Hasnain. 2010. 'Theoretical Analysis of Subwavelength High Contrast Grating Reflectors'. *Optics Express* 18 (16): 16973–88.

Kim, Heungsoo, Kwok Cheung, Raymond C.Y. Auyeung, Donald E. Wilson, Kristin M. Charipar, Alberto Piqué, and Nicholas A. Charipar. 2019. 'VO$_2$-Based Switchable Radiator for Spacecraft Thermal Control'. *Scientific Reports* 9 (1): 1–8.

Kim, Jae Choon, Zongqing Ren, Anil Yuksel, Ercan M. Dede, Prabhakar R. Bandaru, Dan Oh, and Jaeho Lee. 2021. 'Recent Advances in Thermal Metamaterials and Their Future Applications for Electronics Packaging'. *Journal of Electronic Packaging* 143 (1): 10801.

Kitamura, Rei, Laurent Pilon, and Miroslaw Jonasz. 2007. 'Optical Constants of Silica Glass from Extreme Ultraviolet to Far Infrared at near Room Temperature'. *Appl. Opt.* 46 (33): 8118–33.

Ko, Yeong Hwan, Kyu Jin Lee, Fairooz Abdullah Simlan, Neelam Gupta, and Robert Magnusson. 2023. 'Dual Angular Tunability of 2D Ge/ZnSe Notch Filters: Analysis, Experiments, Physics'. *Advanced Optical Materials* 11 (5): 2202390.

Lee, Minjae, Gwansik Kim, Yeongju Jung, Kyung Rok Pyun, Jinwoo Lee, Byung Wook Kim, and Seung Hwan Ko. 2023. 'Photonic Structures in Radiative Cooling'. *Light: Science and Applications* 12 (1): 134.

Li, H. H. 1980. 'Refractive Index of Silicon and Germanium and Its Wavelength and Temperature Derivatives'. *Journal of Physical and Chemical Reference Data* 9 (3): 561–658.

Li, Xiangyu, Joseph Peoples, Zhifeng Huang, Zixuan Zhao, Jun Qiu, and Xiulin Ruan. 2020. 'Full Daytime Sub-Ambient Radiative Cooling in Commercial-like Paints with High Figure of Merit'. *Cell Reports Physical Science* 1 (10): 100221.

Li, Ying, Wei Li, Tiancheng Han, Xu Zheng, Jiaxin Li, Baowen Li, Shanhui Fan, and Cheng-Wei Qiu. 2021. 'Transforming Heat Transfer with Thermal Metamaterials and Devices'. *Nature Reviews Materials* 6 (6): 488–507.

Liu, Kai, Sangwook Lee, Shan Yang, Olivier Delaire, and Junqiao Wu. 2018. 'Recent Progresses on Physics and Applications of Vanadium Dioxide'. *Materials Today* 21 (8): 875–96.

Long, Linshuang, Sydney Taylor, and Liping Wang. 2020. 'Enhanced Infrared Emission by Thermally Switching the Excitation of Magnetic Polariton with Scalable Microstructured VO$_2$ Metasurfaces'. *ACS Photonics* 7 (8): 2219–27.

Magnusson, Robert. 2014. 'Wideband Reflectors with Zero-Contrast Gratings'. *Optics Letters* 39 (15): 4337.

Maier, Stefan A. 2007. *Plasmonics: Fundamentals and Applications.* Vol. 1. Springer.

Marciniak, Magdalena, Artur Broda, Marcin Gębski, Maciej Dems, Jan Muszalski, Andrzej Czerwinski, Jacek Ratajczak, et al. 2020. 'Tuning of Reflection Spectrum of a Monolithic High-Contrast Grating by Variation of Its Spatial Dimensions'. *Optics Express* 28 (14): 20967.

Marciniak, Magdalena, Marcin Gębski, Maciej Dems, Erik Haglund, Anders Larsson, Majid Riaziat, James A. Lott, and Tomasz Czyszanowski. 2016. 'Optimal Parameters of Monolithic High-Contrast Grating Mirrors'. *Optics Letters* 41 (15): 3495.

Ono, Masashi, Kaifeng Chen, Wei Li, and Shanhui Fan. 2018. 'Self-Adaptive Radiative Cooling Based on Phase Change Materials'. *Optics Express* 26 (18): A777.

Pendry, John Brian. 2000. 'Negative Refraction Makes a Perfect Lens'. *Physical Review Letters* 85 (18): 3966.

Pettit, R. B. 1978. 'Evaluation of Portable Optical Property Measurement Equipment for Solar Selective Surfaces'.

Qiao, Pengfei, Weijian Yang, and Connie J. Chang-Hasnain. 2018. 'Recent Advances in High-Contrast Metastructures, Metasurfaces, and Photonic Crystals'. *Advances in Optics and Photonics* 10 (1): 180.

Qiao, Pengfei, Li Zhu, Weng Cho Chew, and Connie J. Chang-Hasnain. 2015. 'Theory and Design of Two-Dimensional High-Contrast-Grating Phased Arrays'. *Optics Express* 23 (19): 24508–24.

Shi, Run, Nan Shen, Jingwei Wang, Weijun Wang, Abbas Amini, Ning Wang, and Chun Cheng. 2019. 'Recent Advances in Fabrication Strategies, Phase Transition Modulation, and Advanced Applications of Vanadium Dioxide'. *Applied Physics Reviews* 6 (1): 11312.

Shin, Dongheok, Gumin Kang, Prince Gupta, Saraswati Behera, Hyungsuk Lee, Augustine M. Urbas, Wounjhang Park, and Kyoungsik Kim. 2018. 'Thermoplasmonic and Photothermal Metamaterials for Solar Energy Applications'. *Advanced Optical Materials* 6 (18): 1800317.

Smith, David R., John B. Pendry, and Mike C. K. Wiltshire. 2004. 'Metamaterials and Negative Refractive Index'. *Science* 305 (5685): 788–92.

Sun, Kai, Wei Xiao, Callum Wheeler, Mirko Simeoni, Alessandro Urbani, Matteo Gaspari, Sandro Mengali, C. H. Kees De Groot, and Otto L. Muskens. 2022. 'VO_2 metasurface Smart Thermal Emitter with High Visual Transparency for Passive Radiative Cooling Regulation in Space and Terrestrial Applications'. *Nanophotonics* 11 (17): 4101–14.

Tang, Kechao, Kaichen Dong, Jiachen Li, Madeleine P. Gordon, Finnegan G. Reichertz, Hyungjin Kim, Yoonsoo Rho, et al. 2021. 'Temperature-Adaptive Radiative Coating for All-Season Household Thermal Regulation'. *Science* 374 (6574): 1504–9.

Taylor, Sydney, Yue Yang, and Liping Wang. 2017. 'Vanadium Dioxide Based Fabry-Perot Emitter for Dynamic Radiative Cooling Applications'. *Journal of Quantitative Spectroscopy and Radiative Transfer* 197: 76–83.

Wang, Hao, Yue Yang, and Liping Wang. 2014. 'Wavelength-Tunable Infrared Metamaterial by Tailoring Magnetic Resonance Condition with VO_2 phase Transition'. *Journal of Applied Physics* 116 (12): 071907.

Wang, Liping, Soumyadipta Basu, and Zhuomin M. Zhang. 2012. 'Direct Measurement of Thermal Emission from a Fabry-Perot Cavity Resonator'. *Journal of Heat Transfer* 134 (7): 072701.

Yang, Chiyu, Preston Bohm, Wenshan Cai, and Zhuomin M. Zhang. 2023. 'Fluctuational Electrodynamics and Thermal Emission'. In *Light, Plasmonics and Particles*, edited by M Pinar Mengüç and Mathieu Francoeur, 43–67. Elsevier.

Yang, Yi-Hua, Jui-Yung Chang, Dong-Han Wu, and Yu-Bin Chen. 2021. 'Impacts from Triple Phases of a Germanium-Antimony-Tellurium Film Coating on Thermal Emission from SiO_2 and Boron Doped Si.' *Optical Materials Express* 11 (9): 3071–3078.

Zhang, Richard Z., and Ken Araki. 2023. 'Ultralow Emittance Thermal Radiation Barrier Achieved by a High-Contrast Grating Coating'. *Journal of Thermophysics and Heat Transfer* 37 (1): 227–39.

Zhang, Wen-Wen, Hong Qi, An-Tai Sun, Ya-Tao Ren, and Jing-Wen Shi. 2020. 'Periodic Trapezoidal VO_2-Ge Multilayer Absorber for Dynamic Radiative Cooling'. *Optics Express* 28 (14): 20609.

Zhang, Xiang, Yanlong Tian, Wenjie Li, Shuliang Dou, Lebin Wang, Huiying Qu, Jiupeng Zhao, and Yao Li. 2019. 'Preparation and Performances of All-Solid-State Variable Infrared Emittance Devices Based on Amorphous and Crystalline WO_3 Electrochromic Thin Films'. *Solar Energy Materials and Solar Cells* 200: 109916.

Zhang, Zhuomin M. 2020. *Nano/Microscale Heat Transfer*. Springer Nature.

Zheludev, Nikolay I., and Yuri S. Kivshar. 2012. 'From Metamaterials to Metadevices'. *Nature Materials* 11 (11): 917–24.

CHAPTER 5

Adaptive Thermal Radiation Control by Si/VO$_2$ Metasurfaces

Junichi Takahara

Osaka University, Osaka, Japan

5.1 INTRODUCTION

Thermal radiation is the spontaneous emission from thermally excited atoms, and the radiation spectrum is determined only by the natural constants according to Planck's law. Although spontaneous emission is considered a natural property of an atom itself, Purcell pointed out that it can be modified depending on the external environment where the atom is placed (1). In particular, when the size of the space where an atom is placed is approximately the wavelength of light (λ), the space becomes a resonator and spontaneous emission is suppressed (decrease in the spontaneous emission rate) or conversely amplified (increase in the spontaneous emission rate) compared to an atom in free space (2, 3). This is known as the cavity quantum electrodynamics effect.

It is natural to assume that the thermal radiation can be controlled by resonators. In 1986, deep diffraction gratings on a solid surface were used to show that the thermal radiation spectrum increased resonantly as the grooves of the gratings became aperture-type resonators (4). This is the first example of thermal radiation control using a microcavity array (MCA). The control of thermal radiation using nanotechnology has a history of more than 30 years since basic studies using MCAs were conducted. Thermal radiation from various types of nano- and microstructures, such as multilayers, photonic crystals, metamaterials, and metasurfaces, has been researched (5).

The advantage of thermal-radiation control is its ability to convert a broad and incoherent thermal radiation spectrum into a specific angular frequency (wavelength) within a narrow band. A perfect absorber (PA) absorbs all light at an angular frequency ω. According to Kirchhoff's law, the relative spectral absorptivity $\alpha(\omega)$ of an object at ω is equal to its relative spectral emissivity $\varepsilon(\omega)$. The emissivity of a PA can be maximized to the blackbody level, that is, $\varepsilon(\omega) = 1$. In principle, a PA is considered a mutual energy converter between light and heat, with 100% conversion efficiency. Hence, a PA can efficiently emit light at specific wavelengths if heated according to Kirchhoff's law. Therefore, in the early stages of research, their applications were limited to narrowband infrared (IR) emitters or thermo-photovoltaic cells.

PAs have been recognized as promising devices for high-efficiency energy conversion. In recent years, their applications have expanded widely, owing to the emergence of new fields such as radiative cooling and thermoelectric power generation (6). In addition, future applications are expected to include energy harvesting, where the IR components of sunlight and thermal radiation from an object at temperatures below 500 K, which have been discarded thus far, are recovered as electric power sources. Thus, thermal radiation control has progressed beyond simple IR emitters, and it is now considered a fundamental technology for a sustainable society.

Recently, Kubo et al. demonstrated that environmental power generation is possible when a plasmonic metasurface is formed on only one side of a thermoelectric material placed at a "uniform" temperature (7). This finding reaffirms the importance of a PA based on a sufficiently thin metasurface compared to λ. Although various types of metasurface structures have been proposed over the past two decades, dynamic control of the thermal radiation spectrum remains challenging. Tunable metasurfaces with suitable optical properties are required for adaptive thermal management.

This chapter reviews our recent studies on static and active PAs based on metasurfaces. We demonstrated refractory plasmonic PAs and hybridized Si/VO$_2$ metasurfaces mediated by the plasmonic and all-dielectric properties of VO$_2$ caused by a metal-insulator transition (MIT). We achieved switchable emissivity and thermal radiation control by heat in the environment for adaptive radiative cooling.

5.2 STATIC PERFECT ABSORBER USING A PLASMONIC METASURFACE

A plasmonic metasurface based on three layers of a metal–dielectric–metal (MDM) structure is currently a promising PA structure for thermal radiation control (8). Figure 5.1(a) shows the schematic of a typical PA based on a metasurface with a patchy plasmonic resonator composed of an MDM structure as the meta-atom.

Plasmonic resonators are optical resonators of surface plasmon polaritons (SPPs), which are the coupled states (polaritons) of light to surface plasmons (SPs). SPPs are the surface mode of plasmons, namely, the collective oscillations of free electrons in a bulk metal. The electromagnetic field of an

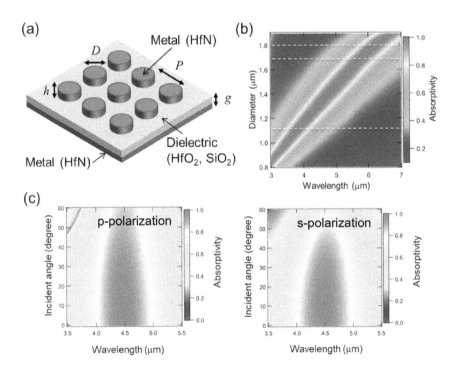

FIGURE 5.1 Structure and design of a static PA based on a refractory metasurface. (a) Schematic of cylindrical MDM meta-atoms with diameter D, dielectric gap thickness g, top metal thickness h, and period P, (b) simulated spectral absorptivity map to D of the metasurface composed of HfN and SiO$_2$ on a quartz substrate with $g = 130$ nm, $h = 200$ nm, and $P = 2.0$ μm (horizontal dotted lines: D of the meta-atoms of Figure 5.3(b)), and (c) simulated spectral absorption map to the incident angle for p- and s-polarizations. (Reprinted with modifications from H. Toyoda, K. Kimino, A. Kawano, and J. Takahara, Photonics, 6(4), 105 (2019). (23).)

SPP exists and propagates at the metal/dielectric interface; hence, it can be strongly confined within a nanogap in an MDM structure as a plasmonic resonator. By reducing the thickness of the resonator, the wavelength of an SPP inside a plasmonic resonator (λ_{SPP}) can be reduced to subwavelength order compared to the vacuum wavelength λ_0 (9, 10). Thus, the size of the resonator can be reduced without cut-off beyond the diffraction limit of light ($\ll \lambda_0$). Because the light incident on the plasmonic resonator is converted to SPPs and dissipates as heat owing to ohmic loss in metals, we can use an MDM structure to realize a narrow-band PA with high efficiency.

MDM-based plasmonic metasurfaces were first theoretically proposed and experimentally studied in 2008 using an IR light source (11, 12). Several studies on IR light sources based on MDM structures were reported during the 2010s (13–16). In addition, a multi-band spectrum IR emitter was realized by combining multiple-sized MDM structures (17). Even a very large-area (approximately 1 m²) IR emitter was demonstrated (18). However, conventional plasmonic metasurfaces are composed of noble metals such as silver (Ag) or gold (Au), which cannot withstand operation at high temperatures above 1000 K.

The recent diversification of plasmonic materials has led to the discovery of negative dielectric (ND) materials with high melting points similar to those of refractory metals such as tungsten (W) but with dielectric constants close to those of noble metals. Metal nitride ceramic materials such as titanium nitride (TiN), zirconium nitride (ZrN), and hafnium nitride (HfN) are collectively referred to as refractory plasmonic materials (19, 20). The melting points and operating wavelength regions of the ND of typical refractory metals and refractory plasmonic materials are summarized in Table 5.1. Metamaterials fabricated using refractory plasmonic materials can be applied to PAs and thermal radiation emitters that can withstand high temperatures. This field is called "refractory plasmonics." Operation at temperatures above 1000 K is now possible, which was previously considered impossible for plasmonics (21, 22).

Melting points and wavelength regions of ND for typical refractory metals and refractory plasmonic materials.

In this study, we designed and demonstrated MDM metasurfaces based on a refractory plasmonic material (23). We used HfN as a refractory plasmonic material because its melting point is above 3300 K, which is equivalent to that of tungsten, as shown in Table 5.1. Figure 5.1(b) and (c) show the simulated spectral absorptivity map obtained by changing the diameter and its angular dependence in a refractory plasmonic metasurface

TABLE 5.1   Melting Points and D/ND Range in Typical Refractory Metals and
Plasmonic Materials

Materials	Melting point (°C)	Dielectric(D)/Negative Dielectric (ND)
Al	660	ND ($\lambda_0 > 140$ nm)
Ag	962	ND ($\lambda_0 > 340$ nm)
Au	1064	ND ($\lambda_0 > 400$ nm)
Quartz (SiO$_2$)	1650	D, ND (8.2 µm $< \lambda_0 <$ 9.4 µm)
TiO$_2$	1870	D
Sapphire (Al$_2$O$_3$)	2072	D
Mo	2623	ND ($\lambda_0 > 850$ nm)
HfO$_2$	2758	D
TiN	2930	ND ($\lambda_0 > 480$ nm)
ZrN	2980	ND ($\lambda_0 > 400$ nm)
Ta	3020	ND ($\lambda_0 > 600$ nm)
HfN	3334	ND ($\lambda_0 > 400$ nm)
W	3422	ND ($\lambda_0 > 940$ nm)

with HfN (ND) disks, SiO$_2$ (dielectric) gap layer, and HfN bottom layer
(ND) on a quartz substrate. Numerical simulations were performed using
commercially available finite-difference time-domain (FDTD) software
(Lumerical). Figure 5.1(b) suggests that the resonant peak wavelength of
the absorptivity can be tuned in the range of 3–7 µm by controlling the
diameter D of the meta-atom. As shown in Figure 5.1(c), the main peak
wavelength of the absorptivity is maintained, even at an incident angle of
40°. Such robustness to angular variation is a typical property of metasurfaces, which differs from conventional diffraction gratings.

Figure 5.2(a) and (b) show sample photographs of an HfN thin-film for
reference and an MDM metasurface with $D = 1.14$ µm, 1.67 µm, and 1.8
µm, respectively (23). Nitrogen gas was not used during fabrication, and
an HfN target was used to deposit the film. The metasurface structures
were fabricated using electron beam lithography and lift-off after thin film
formation.

To investigate the stability of HfN nanostructures at high temperature,
structures with 3 µm squared were fabricated with HfN and Au for comparison, and annealed at 800°C for 15 min. Consequently, the structure of
Au was significantly deformed and destroyed after annealing, whereas no
structural changes were observed in HfN. Therefore, the stability and
durability of the HfN microstructure at high temperatures are superior
to those of Au. We measured the reflectivity spectra, $T(\lambda)$, at room

(a)   (b)   (c)

FIGURE 5.2   Photographs of the samples: (a) HfN thin film (15 × 15 mm) for reference, (b) metasurface (15 × 15 mm) composed of HfN and SiO_2 on a quartz substrate, and (c) SIM image of meta-atoms of (b) with D = 1.14 μm, P = 2.0 μm, g = 130 nm, and h = 200 nm. The scale bar is 1 μm. (Reprinted with modifications from H. Toyoda, K. Kimino, A. Kawano, and J. Takahara, Photonics, 6(4), 105 (2019). (23).)

temperature using microscopic Fourier-transform infrared (FTIR) spectroscopy (VERTEX 70v and HYPERION 2000, Bruker). The absorptivity spectra, $A(\lambda)$, were obtained using $A(\lambda) = 1 - T(\lambda)$.

Figure 5.3(a) shows the calculated and experimental absorption spectra of the HfN thin films at mid-IR wavelengths. We observed low (<0.15) and flat absorptivity spectra with high metallic reflectivity owing to the ND. Slightly higher absorptivity was observed in the experiment than in the simulations. Figure 5.3(b) shows the experimental results of the absorptivity spectra of the MDM metasurfaces with three diameters: D = 1.14, 1.67, and 1.8 μm. We observe that the peak wavelengths change from 4 μm to 6.5 μm as D increases. The horizontal dotted lines in Figure 5.1(b) confirmed that the resonant peaks in the absorptivity could be attributed to plasmonic resonances.

Such MDM metasurfaces are PA in theory followed by the simulated results in Figure 5.1. In the experiment, considering the case of D = 1.14 μm as an example, we observed an absorption peak at approximately 4 μm, which was consistent with the simulated value. However, the experimental absorptivity of approximately 0.84 at the peak wavelength was approximately 10–15% lower than the simulated values. This is because the Cassegrain IR objective lens used in microscopic FTIR cannot measure the absorptivity in the normal direction, but only at an incidence angle of 10–23°. In the case of D = 1.14 μm, the simulated and experimental Q-values were Q = 2.9 and 1.7, respectively. Although the actual values

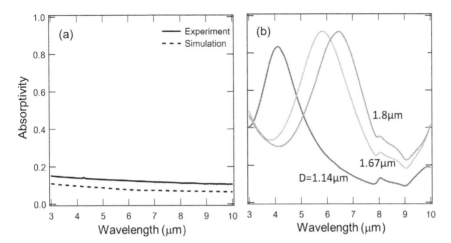

FIGURE 5.3 Absorption spectra of the MDM metasurface based on HfN. (a) Measured (solid line) and simulated (dotted line) absorption spectra of HfN thin film shown in Figure 5.2(a) and (b) measured absorption spectra of the MDM metasurfaces composed of HfN and SiO_2 on a quartz substrate shown in Figure 5.2(b) with $D = 1.14$ μm, 1.67 μm, and 1.8 μm ($P = 2.0$ μm, $g = 130$ nm, and $h = 200$ nm).

are expected to be closer to the simulations for the above reasons, the Q-values in the plasmonic metasurface are only of the order of approximately 1. The low Q-value is a shortcoming of plasmonic metasurfaces.

5.3 METAL-INSULATOR TRANSITION AND OPTICAL PROPERTIES OF VO₂

Since the 2000s, active metamaterials that dynamically change their optical properties via changes in the structure or refractive index of a meta-atom have been proposed (24, 25). A meta-atom that includes phase transition materials is promising for active metamaterials and has attracted much attention because no moving parts are used and the optical properties can be reversibly changed (26). Ge-Sb-Te (GST) and vanadium dioxide (VO_2) are well-known phase transition materials. GST is a chalcogenide compound, also known as an optical disc material. VO_2 is a transition metal oxide with strongly correlated electrons that exhibits MIT, in which it becomes a Mott insulator at room temperature and a metal above the MIT temperature (T_{MI}) (27, 28). While the glass transition temperature (T_g) of GST is approximately 600°C (873 K), the T_{MI} of VO_2 is as low as 68°C (341 K). Because the T_{MI} of VO_2 is much lower than the T_g of GST, it is suitable for applications in adaptive thermal management. In addition, T_{MI}

can be systematically reduced by doping with W^{6+} or other materials (28, 29). Hence, VO_2 can be applied for the adaptive control of thermal radiation, in which the emissivity varies over a wide range of temperatures.

Figure 5.4 shows the change in the crystal structure of VO_2 and temperature dependence of the resistivity (27, 30). As shown in Figure 5.4(a), the crystal structure of VO_2 is orthorhombic (monoclinic) below T_{MI} and tetragonal (rutile-type) above it (27). Vanadium and atoms on the (200) plane of VO_2 are shown in Figure 5.4(a). When the temperature is varied above and below T_{MI}, the position of the vanadium atoms shifts and changes between the orthorhombic and tetragonal structures occur,

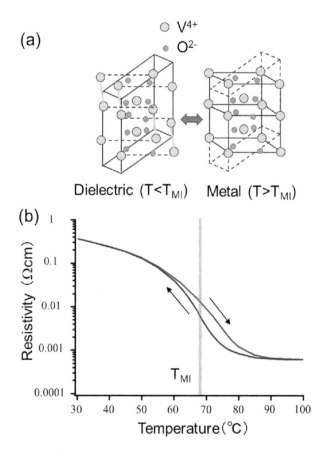

FIGURE 5.4   Crystal structures and resistivity of VO_2. (a) Monoclinic insulator (dielectric) at $T < T_{MI}$ and rutile metal at $T > T_{MI}$ and (b) temperature dependence of resistivity. The arrows indicate hysteresis. (Reprinted with modification in permission from J. Takahara, KOGAKU 52(7) 274 (2023). Copyright 2023, The Optical Society of Japan (30).)

resulting in an energy shift in the band structure. This band shift causes an MIT, resulting in a significant change in resistivity. Figure 5.4(b) shows that the resistivity changes by approximately three orders of magnitude with hysteresis because of the MIT. The permittivity of VO$_2$ also changes with the MIT. If VO$_2$ is included as a part of the metasurface, the optical properties of the metasurface can be tuned significantly by varying the temperature.

To measure the optical properties of VO$_2$, we prepared VO$_2$ thin films on both Si and sapphire (Al$_2$O$_3$) substrates using pulsed laser deposition (PLD), as described below. First, a VO$_2$ thin film with a thickness of 400 nm was prepared using PLD. The target was V$_2$O$_5$, the most stable vanadium oxide, and its composition is not easily changed by oxidation in air. The substrate was a (0001) surface of Al$_2$O$_3$ (c-plane sapphire) whose lattice constant matched that of the (010) surface of VO$_2$. In addition, a c-plane sapphire substrate is suitable because the thermal conductivity of the substrate must be high during heating. X-ray diffraction measurements of a sample of VO$_2$ deposited on a sapphire substrate showed that in addition to the peaks arising from the c-plane sapphire substrate, peaks in the (020) and (040) planes of VO$_2$ were observed. This confirmed that c-axis-oriented VO$_2$ crystals were deposited on the sapphire substrate.

The VO$_2$-deposited sapphire substrate was placed on a polyimide heater and the complex permittivity of the VO$_2$ thin film was measured by phase-modulation ellipsometry. Here, the permittivity was measured at temperatures of $T = 23°C$ ($T < T_{MI}$) (dielectric phase) and $T = 75°C$ ($T > T_{MI}$) (metallic phase). Figure 5.5 shows the measured near-IR spectra of the complex relative permittivity of VO$_2$ at these temperatures. We observed that the real part of the permittivity had positive values (>0) in the dielectric phase and negative values (<0) in the metallic phase 31). The negative value of the permittivity is the evidence of the metallic phase of VO$_2$ at $T = 75°C$. As the metallic phase of VO$_2$ exhibits ND only at $\lambda_0 > 1100$ nm, it can be used as a transmissive optical material in the visible region. The sign of the permittivity switched reversibly between positive and negative values owing to the MIT by controlling the temperature.

Figure 5.6 shows the temperature dependence of the reflectivity R measured at 10 μm in the VO$_2$ thin film ((refractive index $n\sim2.8$) formed on Si ($n\sim3.5$) and Al$_2$O$_3$ substrate ($n\sim0.96$). In both cases, the reflectivity rapidly changes with the temperature owing to the MIT. In the dielectric phase at $T < T_{MI}$, R is relatively low ($R\sim0.35$) in the Si substrate and $R < 0.1$ in the Al$_2$O$_3$ substrate. In the metallic phase at $T > T_{MI}$, the reflectivity is high

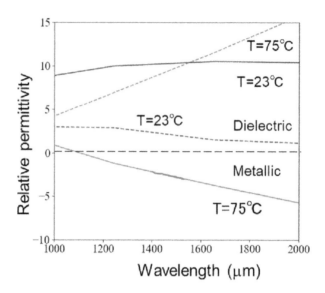

FIGURE 5.5 Real (solid lines) and imaginary part (dotted lines) of relative permittivity spectra of VO$_2$ thin film measured in near-IR region at $T = 23°C$ and 75°C. (Reprinted with modification in permission from J. Takahara, KOGAKU 52(7) 274 (2023). Copyright 2023, The Optical Society of Japan (30).)

($R = 0.8$–0.9) in both substrates, which is consistent with the metallic phase of VO$_2$. Hysteresis is observed only for the Al$_2$O$_3$ substrate. We used the Al$_2$O$_3$ substrate instead of the Si substrate because of the high contrast in the reflectance change.

5.4 SWITCHABLE PERFECT ABSORBER USING A VO$_2$ METASURFACE

Figure 5.7 shows the structures of active metasurfaces that are partly composed of VO$_2$ (30). We achieved reversible switching of absorptivity (emissivity) in a VO$_2$ metasurface with high contrast by changing the temperature as a parameter. We define the following states of emissivity: the ON state (higher emissivity) as $\alpha(\omega) = \varepsilon(\omega) > 0.9$, and the OFF state (lower emissivity) as $\alpha(\omega) = \varepsilon(\omega) < 0.1$.

As shown in Figure 5.7(a), meta-atom disks placed on a dielectric (Al$_2$O$_3$) substrate are composed of VO$_2$. At temperatures below T_{MI} ($T < T_{MI}$), the metasurface is all-dielectric because VO$_2$ is a dielectric, and the meta-atom is a dielectric Mie resonator. Above T_{MI} ($T > T_{MI}$), the metasurface transforms into plasmonic because VO$_2$ changes from dielectric to metal, and the meta-atom is a plasmonic resonator. This MIT shifts the

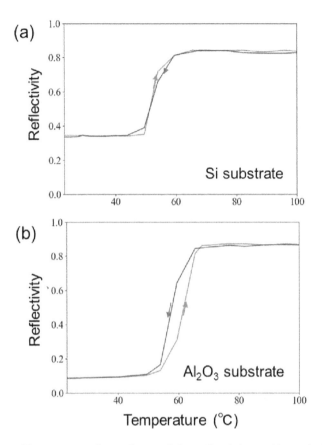

FIGURE 5.6 Temperature dependence of the reflectivity at 10 μm in VO₂ thin films formed on (a) Si and (b) Al₂O₃ substrates. The arrows indicate hysteresis.

resonant wavelength of the metasurface, resulting in a slight change in the reflectivity (24, 25). However, this structure is unsuitable for practical use because the reflectivity is approximately 0.1–0.2 owing to the intrinsic loss of VO₂, and the contrast of the reflectivity change caused by the phase transition is low (31).

The structures of Figure 5.7(b) and (c) are complementary; both are plasmonic metasurfaces based on MDM structures. Figure 5.7(b) illustrates a metal meta-atom formed on a metal substrate with a VO₂ gap layer (27). Because this is a typical MDM structure at $T < T_{MI}$, perfect absorption occurs. Thermal radiation becomes ON ($\varepsilon(\omega) \sim 1$) at the resonant wavelength and turns OFF ($\varepsilon(\omega) \sim 0.1$) at $T > T_{MI}$ because of the metallic nature of the entire structure. Figure 5.7(c) shows VO₂ meta-atoms formed on a metal substrate with a dielectric gap layer (26). In contrast to Figure 5.7(b),

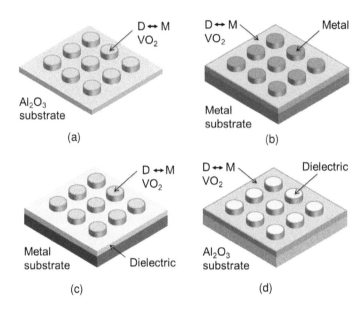

FIGURE 5.7   Structures of the active metasurface partially including VO_2. (a) VO_2 disks on a dielectric (Al_2O_3) substrate, (b) metal disks on a metal substrate with a VO_2 gap layer, (c) VO_2 disks on a metal substrate with a dielectric gap layer, and (d) dielectric disks on a dielectric (Al_2O_3) substrate with a VO_2 gap layer. VO_2 mutually changes between dielectric (D) and metal (M) by MIT. (Reprinted with modification in permission from J. Takahara, KOGAKU 52(7) 274 (2023). Copyright 2023, The Optical Society of Japan (30).)

the thermal radiation is OFF at $T < T_{MI}$ and ON at $T > T_{MI}$. Because this structure changes to an MDM structure at $T > T_{MI}$, perfect absorption occurs and thermal radiation is in the ON state. Hence, these structures are suitable for the adaptive control of radiative cooling. However, their applications are limited because these structures are opaque due to a metal layer at the bottom. On an opaque substrate in the visible range, sunlight cannot be used during the daytime.

In contrast, the structure is transparent in the visible range, as shown in Figure 5.7(d) (32). This structure is the so-called "hybridized metasurface" with dielectric/plasmonic structures. Figure 5.7(d) shows dielectric Mie resonators on a dielectric (Al_2O_3) substrate with a VO_2 gap layer (28, 29). Because a dielectric Mie resonator has a higher Q-value value than a plasmonic resonator, such hybridized metasurfaces are expected to have higher Q-values in the ON state. At $T < T_{MI}$, the metasurface is all-dielectric and transmits visible and IR light, as shown in Figure 5.5. At $T > T_{MI}$, it becomes absorptive at the resonant wavelength, owing to the significant loss of plasmonic absorption in the Mie resonator on the metal layer. Thermal radiation is OFF at $T < T_{MI}$ and ON at $T > T_{MI}$.

Adaptive control of emissivity according to the atmospheric temperature is required, especially for radiative cooling, because the cooling effect must be turned off when the temperature decreases at night (33, 34). Hence, the aforementioned switchable PA structures can be applied to the adaptive radiative cooling of roofs, windows, and solar cells.

5.5 SWITCHABLE THERMAL RADIATION CONTROL USING AN SI/VO₂ METASURFACE

We demonstrated a hybridized metasurface for adaptive radiative cooling, as shown in Figure 5.8(a), which is an implementation of that shown in Figure 5.7(d), using Si as a dielectric Mie resonator. Figure 5.8(b) shows a photograph of the sample and scanning ion microscope (SIM) image of the fabricated meta-atoms.

Figure 5.9 shows the fabrication process of the metasurface. First, a VO_2 thin film with a thickness of 300 nm was deposited using PLD. The substrate was a (0001) surface of Al_2O_3 (c-plane sapphire) whose lattice constant matched that of the (010) surface of VO_2. Then, amorphous Si (a-Si)

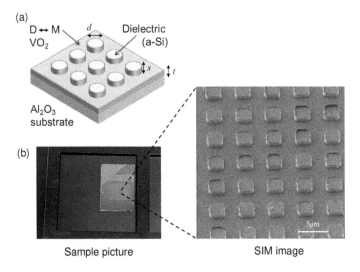

FIGURE 5.8 Switchable PA using Si/VO₂ metasurface. (a) Schematic of the cylindrical Si/VO₂ hybridized meta-atoms with a diameter of d, VO₂ gap thickness t, top Si thickness s, and period P, (b) picture of the sample and SIM image of meta-atoms, where $d = 2.4$ μm, $t = 300$ nm, $s = 800$ nm, and $P = 5$ μm. The scale bar is 5 μm.

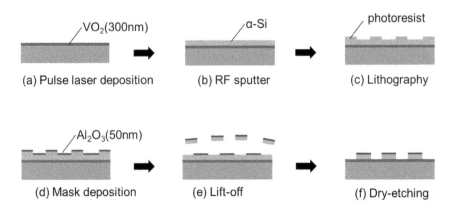

FIGURE 5.9 Fabrication process of the Si/VO$_2$ metasurface. (a) VO$_2$ film deposition (300 nm) by PLD, (b) a-Si film deposition (800 nm) by RF sputtering, (c) spin-coating photoresist OFPR800LB (1000 nm) and mask-less lithography, (d) Al$_2$O$_3$ mask deposition (50 nm) by RF sputtering, (e) lift-off to remove the resist, and (f) RIE of a-Si layer.

was deposited with a thickness of 800 nm on the VO$_2$ thin film by RF sputtering (SVC-700LRF, Sanyu Electron Co., Ltd.). A maskless lithography system (D-Light DL-1000, NanoSystem Solutions, Inc.) was used to pattern the metasurface structure. We then transferred the pattern to a-Si by depositing a 50-nm-thick Al$_2$O$_3$ mask by RF sputtering and reactive-ion etching (RIE) (ELIONIX Inc.). Finally, we obtained the sample of the Si/VO$_2$ metasurface, with meta-atom diameter of $d = 2.4$ μm, thickness $h = 800$ nm, VO$_2$ film thickness $h_2 = 300$ nm, and period $P = 5$ μm.

To measure the switching of absorptivity, we set the sample on a ceramic heater, heated it in air, and measured the reflectivity spectra $T(\lambda)$ by microscopic FT-IR (VERTEX 70v and HYPERION 2000, Bruker). The absorptivity spectra, $A(\lambda)$, were obtained using $A(\lambda) = 1-T(\lambda)$. Figure 5.10(a) shows the changes in the mid-IR absorptivity spectra caused by the MIT (32). We observed low (~0.1) and relatively flat absorptivity spectra in 5–13 μm at $T = 23°C$, which is consistent with the all-dielectric phase of Figure 5.8(a). When T is 100°C, that is, above T_{MI}, we observed resonant peaks in absorptivity in the 5–6 μm and 8–13 μm regions, which correspond to the atmospheric windows.

To clarify the origin of the resonance, we performed FDTD simulations (Lumerical). In Figure 5.10(a), the dotted lines show the results of the FDTD simulations. The experimental and simulated results were in good quantitative agreement. The theoretical value of the intensity ratio between

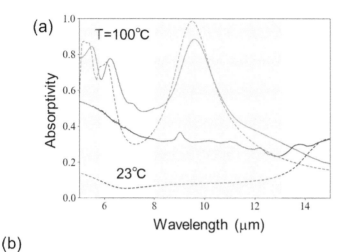

(a)

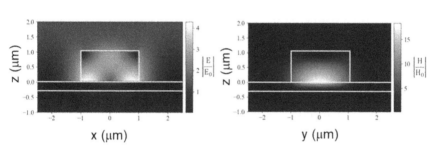

(b)

FIGURE 5.10 Absorptivity of the Si/VO$_2$ metasurface. (a) Simulated (dotted lines) and measured (solid lines) absorptivity spectra in mid-IR region at $T = 23°C$ and $100°C$ (shaded areas: atmospheric windows. (32)), (b) normalized electric (left) and magnetic (right) fields by incident filed strength of the metallic phase at $T = 100°C$.

the ON/OFF states of approximately 10 at 9.8 μm, which is greater than that of previously reported plasmonic metasurfaces. The main peak at approximately 9.8 μm was attributed to the magnetic dipole resonance in the dielectric Mie resonator. This is because the electric dipole resonance is canceled owing to the mirror effect of the substrate in the ON state. Figure 5.10(b) suggests that electric- and magnetic-field enhancements occur at the interface of a-Si/VO$_2$. These switching effects indicate that the emissivity can be switched between the ON and OFF states according to the temperature.

Next, we directly measured the switching characteristics of the thermal radiation. For the measurements, all the optical systems were set in a vacuum chamber connected to an FT-IR spectrometer (FT/IR 6000,

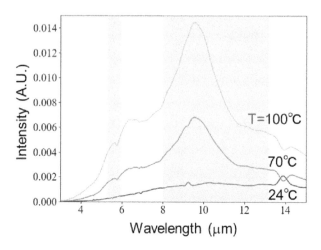

FIGURE 5.11   Thermal radiation spectra in the mid-IR region at T = 24, 70, and 100°C (from bottom to top). The shaded areas represent the atmospheric windows. (32)

JASCO Co.) through a tunnel tube to prevent oxidation. The vacuum chamber and spectrometer were pumped at 2.0×10^2 Pa and 1.4×10^2 Pa, respectively. The spectral resolution was set at 4 cm^{-1} and a DLATGS detector was used for the measurement. The samples were placed on a microceramic heater (MS-1000, Sakaguchi E.H. VOC Corp.). The temperature of the sample was measured using a K-type Sheath thermocouple (T350251H, Sakaguchi E.H. VOC Corp.) placed on its surface. The measurements were performed at 23°C and 100°C.

Figure 5.11 shows the far-field thermal radiation spectra measured under vacuum conditions (32). When the temperature was increased above T = 70°C, the radiation intensity was significantly enhanced, with a peak around 10 µm, corresponding to the atmospheric window. Thus, we experimentally achieved the switching of the thermal radiation intensity by the MIT of VO_2, as predicted from the absorptivity. Because the MIT temperature can be reduced by doping, we can tune the radiative cooling effects and achieve switching at the optimum temperature in the future (27, 28).

5.6 SUMMARY

We have reviewed PAs based on metasurfaces and demonstrated the switchable control of thermal radiation using an Si/VO_2 hybridized metasurface. We reversibly changed the emissivity of the metasurface between ON and OFF states by varying the temperature.

Recently, optical switching using the thermo-optic effects of PA by the Huygens dipole has been achieved in Si metasurfaces and is expected to be applied to all-optical switching devices (35). Because a structure thinner than the wavelength of the metasurface is essential in regions where the effect of thermal conduction is large, active metasurface PAs mediated by VO_2 are expected to be key devices in the fields of thermo-optic effects and thermoelectric conversion in the future.

ACKNOWLEDGMENTS

I would like to thank Hirofumi Toyoda, Toshinori Kohno, and Hiroaki Takase for their great efforts in numerical simulations and experiments during their master's studies. I thank Dr. Yusuke Nagasaki and Kazunari Kimino for their support in sample fabrication and measurements. We thank the Nanotechnology Open Facility (NOF) at the Institute of Scientific and Industrial Research (SANKEN), Osaka University for nano-fabrication by MEXT: "Nanotechnology Platform Project" [F-20-OS-0020] [S-19-OS-0003]. This study was partly supported by the Japan Society for the Promotion of Science (JSPS) KAKENHI (19H02630), a research grant from The Murata Science Foundation, and joint research funds from Sumitomo Electric Industries, Ltd.

REFERENCES

1. E.M. Purcell, *Phys. Rev.* 69 (1946) 681, https://doi.org/10.1103/PhysRev.69.37
2. R.G. Hulet, E.S. Hilfer, and D. Kleppner, *Phys. Rev. Lett.* 55(20) (1985) 2137, https://doi.org/10.1103/PhysRevLett.55.2137
3. D.J. Heinzen, J.J. Childs, J.E. Thomas, and M.S. Feld, *Phys. Rev. Lett.* 58(13) (1987) 1320, https://doi.org/10.1103/PhysRevLett.58.1320
4. P.J. Hesketh, J. N. Zemel, and B. Gebhart, *Nature* 324 (1986) 549, https://doi.org/10.1038/324549a0
5. D.G. Baranov, Y. Xiao, I.A. Nechepurenko, A. Krasnok, A. Alu and M.A. Kats, *Nature Mat.* 18 (2019) 920, https://doi.org/10.1038/s41563-019-0363-y
6. K.T. Lin, J. Han, K. Li, C. Guo, H. Lin, and B. Jia, *Nano Energy* 80 (2021) 105517, https://doi.org/10.1016/j.nanoen.2020.105517
7. S. Katsumata, T. Tanaka, and W. Kubo, *Opt. Express* 29 (2021) 16396, https://doi.org/10.1364/OE.418814
8. S. Ogawa and M. Kimata, *Materials* 11 (2018) 458, https://doi.org/10.3390/ma11030458
9. J. Takahara, S. Yamagishi, H. Taki, A. Morimoto and T. Kobayashi, *Opt. Lett.*, 22 (1997) 475, https://doi.org/10.1364/OL.22.000475
10. J. Takahara and T. Kobayashi, *Optics & Photonics News*, 15, No. 10, (2004) 54, https://doi.org/10.1364/OPN.15.10.000054

11. B.J. Lee, L.P. Wang and Z.M. Zhang, *Opt. Express* 16(15) (2008) 11328, https://doi.org/10.1364/OE.16.011328

12. I. Puscasu, W.L. Schaich, *Appl. Phys. Lett.* 92 (2008) 233102, https://doi.org/10.1063/1.2938716

13. X. Liu, T. Tyler, T. Starr, A. F. Starr, N. M. Jokerst, and W. J. Padilla, *Phys. Rev. Lett.* 107 (2011) 045901, https://doi.org/10.1103/PhysRevLett.107.045901

14. Y. Ueba and J. Takahara, *APEX* 5 (2012) 122001, https://doi.org/10.1143/APEX.5.122001

15. K. Ito, H. Toshiyoshi, and H. Iizuka, *Opt. Express* 24(12) (2016) 12803, https://doi.org/10.1364/OE.24.012803

16. T. Yokoyama, T. D. Dao, K. Chen, S. Ishii, R. P. Sugavaneshwar, M. Kitajima, T. Nagao, *Adv. Opt. Mat.* 4 (12) (2016) 1987, https://doi.org/10.1002/adom.201600455

17. H. Miyazaki, T. Kasaya, M. Iwanaga, B. Choi, Y. Sugimoto and K. Sakoda, *Appl. Phys. Lett.* 105 (2014) 121107, https://doi.org/10.1088/1468-6996/16/3/035005

18. Y. Matsuno and A. Sakurai, *Opt. Mat. Express* 7, 2 (2017) 618, https://doi.org/10.1364/OME.7.000618

19. U. Guler, A. Boltasseva, and V. M. Shalaev, *Science* 344 (2014) 263, https://doi.org/10.1126/science.1252722

20. A. Boltasseva and H.A. Atwater, *Science* 331 (2011) 290, https://doi.org/10.1126/science.1198258

21. G.V. Naik, V.M. Shalaev, and A. Boltasseva, *Adv. Materials* 25 (2013) 3264, https://doi.org/10.1002/adma.201370156

22. M. Kumar, N. Umezawa, S. Ishii, and T. Nagao, *ACS Photon.* 3 (2015) 43, https://doi.org/10.1021/acsphotonics.5b00409

23. H. Toyoda, K. Kimino, A. Kawano, and J. Takahara, *Photonics* 6 (2019) 105, https://doi.org/10.3390/photonics6040105

24. Q. He, S. Sun, and L. Zhou, *Research* 2019 (2019) 1849272, https://doi.org/10.34133/2019/1849272

25. A.M. Shaltout, V.M. Shalaev, and M.L. Brongersma, *Science* 364 (2019) 648, https://doi.org/10.1126/science.aat3100

26. X. Duan, S.T. White, Y. Cui, F. Neubrech, Y. Gao, R.F. Haglund, and N. Liu, *ACS Photon.* 7 (2020) 2958, https://doi.org/10.1021/acsphotonics.0c01241

27. D. Wegkamp and J. Stahler, *Progress Surf. Sci.* 90 (2015) 464, https://doi.org/10.1016/j.progsurf.2015.10.001

28. Z. Shao, X. Cao, H. Luo, and P. Jin, *NPG Asia Materials* 10 (2018) 581, https://doi.org/10.1038/s41427-018-0061-2

29. K. Shibuya, M. Kawasaki, and Y. Tokura, *Appl. Phys. Lett.* 96 (2010) 022102, https://doi.org/10.1063/1.3291053

30. J. Takahara, *KOGAKU* 52(7) (2023) 274.

31. Y. Nagasaki, T. Kohno, K. Bando, H. Takase, K. Fujita, and J. Takahara, *Appl. Phys. Lett.* 115 (2019) 161105, https://doi.org/10.1063/1.5109460

32. H. Takase and J. Takahara, The 68th JSAP Spring Meeting, 17a-Z05-7, (March 17, 2021).

33. A. P. Raman, M. A. Anoma, L. Zhu, E. Rephaeli, and S. Fan, *Nature* 515 (2014) 540, https://doi.org/10.1038/nature13883
34. Z. Chen, L. Zhu, A. Raman, and S. Fan, *Nat. Commun.* 7 (2016) 13729, https://doi.org/10.1038/ncomms13729
35. K. Nishida, K. Sasai, R. Xu, T-H. Yen, Y-L. Tang, J. Takahara, S-W. Chu, *Nanophotonics* 12(1) (2023) 139, https://doi.org/10.1515/nanoph-2022-0597

CHAPTER 6

Metamaterial Thermoelectric Conversion

Wakana Kubo

Koganei, Tokyo, Japan

6.1 INTRODUCTION

Recycling waste heat is essential for creating an energy-efficient society. This is because recycling waste heat leads to a reduction in the use of fossil fuels, which ultimately leads to a decarbonized society. Figure 6.1 presents the ratio of waste heat against the primary energy created from fuels, natural gas, coal, and renewable energy in Japan in 2018. Figure 6.1 presents that waste heat accounts for more than 60% of the ultimate form of energy utilization. In fact, only 40% of the primary energy can be utilized effectively as power, electricity, heat, and light, while the remaining 60% is lost as waste heat. Waste heat is created mainly during the energy purification and conversion process. A similar tendency was observed in United States [1]. Consequently, recycling the waste heat emerges as an urgent issue for realizing a decarbonized society.

Thermoelectric conversion is one of the potential technologies to recycle waste heat because it converts heat into electricity. Thermoelectric devices can therefore be considered to recover waste heat and convert it into a usable energy resource [2–6]. The Seebeck effect plays a significant role in thermoelectric conversion, in which the amplitude of the temperature gradient across a thermoelectric device is the key to determine the electric voltage generated across it. Since the Seebeck effect converts the temperature gradient across a thermoelectric device to electric voltage,

DOI: 10.1201/9781003409090-6

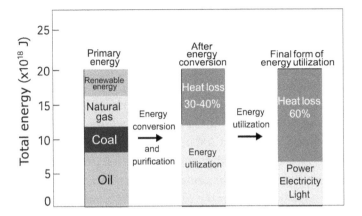

FIGURE 6.1 Proportion of waste heat in the primary energy, energy purification processes, and final form of energy utilization. (Figure 6.1 was independently created based on a report by New Energy and Industrial Technology Development Organization (NEDO) in Japan. (https://www.nedo.go.jp/content/100902074.pdf, written in Japanese. Publication year: 2019).

such devices are usually installed only at the locations where large thermal gradients can be expected, such as space probes and wearable electronics, and the installation sites for conventional thermoelectric devices are strongly limited [7, 8]. Although the Seebeck effect cannot be used under an environment with uniform temperature and thermal radiation, there are a lot of such places, for instance, in water, on the road, inside the furnace, etc. Therefore, new techniques that can create temperature gradients even if the devices are surrounded by environments with uniform thermal radiation are in demand. Such technologies can expand the range of installation of thermoelectric devices, thereby contributing to an energy-recycling and energy-saving society.

In this chapter, we introduce a thermoelectric device loaded with a metamaterial absorber (MA) at the surface of one of its ends to generate electricity even in an environment with uniform thermal radiation [9]. Metamaterials are artificial materials that exhibit extraordinal interaction with electromagnetic waves [10–12]. Using the large difference in absorptivity between the device edges produced by the metamaterial absorber, the device creates a temperature gradient even if the device is basking in uniform and isotropic thermal radiations. Our proposal can break the notion that thermoelectric conversion cannot occur in an environment with uniform thermal radiation. Unlike conventional thermoelectric devices, the metamaterial thermoelectric device can collect and extract thermal energies from the surrounding medium.

Thus, these devices will pave the way to recover waste heat existing in the medium, such as air and water.

6.2 EXPERIMENTAL PROCEDURES

A metamaterial absorber (MA) consisting of a 150 nm-thick Ag film and a 100 nm-thick Ag disk array sandwiching a 60 nm-thick calcium fluoride (CaF_2) layer was fabricated on a Cu electrode whose width, length, and thickness are 4 mm, 6 mm, and 300 μm, respectively (Figure 6.2(a, b)). The diameter and pitch of the Ag disc were 1.8 and 3.0 μm, respectively. The MA arrays were fabricated at the center of the Cu electrode with an area of 2.1 × 2.1 mm² (490,000 units of the MA in the fabricated area).

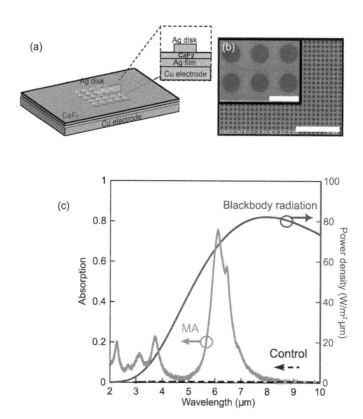

FIGURE 6.2 (a) Schematic of MA arrays fabricated on a Cu electrode. (b) A top-view scanning electron microscope (SEM) image of MA arrays. The inset is a magnified image, and the scale bars in the low and high magnified images represent 30 and 3 μm, respectively. (c) Comparison of an absorption spectra of the MA, a measured absorption spectrum of the control electrode (dashed line), and a calculated blackbody radiation spectrum calculated at 364 K.

Figure 6.2(c) shows the measured absorption spectra of MA and a blackbody radiation at 364 K. The measured absorption spectrum shows a resonance peak at 6.08 μm attributed to its magnetic resonance mode [13], indicating that the MA arrays absorb a part of the thermal radiation emitted from the surrounding medium at 364 K and convert the absorbed energy into local heating [14–23].

The MA-fabricated-copper electrode was attached to one end of a p-type bismuth antimony telluride ($Bi_{0.3}Sb_{1.7}Te_3$), provided by TOSHIMA Manufacturing Co., Ltd, to obtain the thermoelectric device. $Bi_{0.3}Sb_{1.7}Te_3$ was utilized in this research since this material exhibits promising thermoelectric performance. The cross-sectional area and length of the $Bi_{0.3}Sb_{1.7}Te_3$ element were 1×2 mm^2 and 8.8 mm, respectively. A control electrode was attached to the opposite edge of the thermoelectric device. The control electrode consisted of 60 nm-thick CaF_2 and 150 nm-thick Ag layers deposited on a copper electrode.

The dashed line in Figure 6.2(c) is the measured absorption spectrum of the control electrode, indicating that the control electrode does not absorb thermal radiation. The device loaded with the MA fabricated copper electrode is called the MA thermoelectric device (Figure 6.3(a)).

In addition, we prepared a control thermoelectric device loaded with control electrodes at both edges of $Bi_{0.3}Sb_{1.7}Te_3$. To examine the effect of the MA electrode on the thermoelectric performance, we replaced the control electrode of the control device with the MA electrode to model the MA device by retaining the $Bi_{0.3}Sb_{1.7}Te_3$ thermoelectric element of the control device; hence, the difference between the MA device and the control device is only an attachment of the electrode.

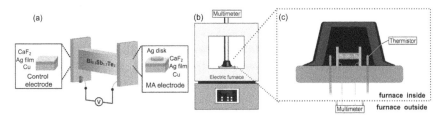

FIGURE 6.3 (a) Schematic of the $Bi_{0.3}Sb_{1.7}Te_3$ thermoelectric device loaded with the MA electrode. The MA electrode was mounted on one end of the $Bi_{0.3}Sb_{1.7}Te_3$ element, and a control electrode was attached on the other end of the element. (b) Experimental setup for thermoelectric measurements using an electric furnace. (c) Schematic of the MA thermoelectric device and two thermistors capped with a carbon pod.

The MA device was fixed on the center of a printed circuit board via a Kapton double-side tape; the substrate was placed in the chamber of an electric furnace (FUL210FA, Advantec). Gold wires (diameter = 0.1 mm) were connected to both electrodes by the Ag paste to create an electrical connection. The gold wires were also connected to Cu wires (diameters = 2 mm), and the Cu wires were connected to the multimeter (DMM-6500, Keithley Instruments), which was placed outside the furnace (Figure 6.3(b, c)). To prevent any impact on the thermoelectric performance, we ensured that the length of the gold and Cu wires connected to both electrodes were the same. The nonuniformity in device structure, arrangement, and wiring can affect the thermoelectric properties. Therefore, we paid close attention to the device arrangement and length of the wiring in our experiment. Figures 6.3(b, c) present the experimental setup for the thermoelectric measurement. To eliminate the convection effect on thermoelectric generation, the MA device was capped with a carbon pod with a diameter and a height of 2.5 and 2.5 cm, respectively (Figure 6.3(c)).

The output voltage across the device was measured using a multimeter. To obtain the output voltage, we averaged the output voltage from 10800 s (180 min) to 11400 s (190 min) after the furnace heating was started. Furthermore, the output voltages were examined with several MA and control electrodes of several different thermoelectrical devices to calculate the average voltages, which indicated that these output voltages were highly reproducible. Because the $Bi_{0.3}Sb_{1.7}Te_3$ element of the MA device was identical to that of the control device, output voltage measurements of these two devices were carried out separately.

6.3 THERMOELECTRIC CONVERSION IN AN ENVIRONMENT WITH UNIFORM THERMAL RADIATION

Figure 6.4(a) illustrates the dependence of the output voltage on the measured environmental temperatures. The values of output voltage generated across the MA and control device were 18.9 ± 6.7 μV and −1.2 ± 4.4 μV (number of the measured sample: 5), respectively, at the measured environmental temperature of 364 K. The output voltage generated across the MA device was significantly larger than that across the control device. The output voltage generated across the MA device corresponded to an additional temperature gradient of 0.14 K estimated by the measured Seebeck coefficient of the p-type $Bi_{0.3}Sb_{1.7}Te_3$ element (−140 μV/K). As we mentioned above, we paid close attention to the device arrangement and length of the wiring in our experiment. We concluded that our control

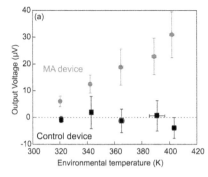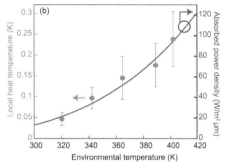

FIGURE 6.4 (a) Dependence of the output voltages generated on the MA device (circle) and a control device (square) on the measured environment temperatures. (b) Correlation between the local heat temperatures generated on the MA-$Bi_{0.3}Sb_{1.7}Te_3$ thermoelectric device and the calculated power density absorbed by the MA.

over the device arrangement and wiring was appropriate because the control device did not exhibit significant thermoelectric properties in an environment with uniform thermal radiation, as shown in Figure 6.4(a). This result indicates that our experimental setup, including a control thermoelectric device, a carbon pod, an electric furnace, and any other materials, does not generate an output voltage. It further confirms that the MA is the key factor responsible for generating electricity in an environment with uniform thermal radiation.

In addition, the output voltages generated across these devices were measured at different environmental temperatures (Figure 6.4(a)). The output voltages generated across the MA thermoelectric device showed significant temperature dependence, whereas those generated on the control device showed no temperature dependence. Figures 6.4(a) shows that the energy absorbed by the MA increases as the environmental temperature increases, indicating an increase in the amount of local heating.

Figure 6.4(b) shows a comparison between the additional temperature gradients estimated by the measured output voltages and the calculated power densities absorbed by the MA. This indicates that the additional temperature gradients created by the MA electrode originate from thermal radiation absorption by the MA.

These results indicate the mechanism of metamaterial thermoelectric voltage generation in an environment with uniform thermal radiation. MA absorbs thermal radiation emitted from the surrounding environment and generates local heating due to absorption losses by the MA. The local heat propagates to the $Bi_{0.3}Sb_{1.7}Te_3$ thermoelectric device via a Cu electrode, resulting in an additional thermal gradient across the $Bi_{0.3}Sb_{1.7}Te_3$ element and subsequent output voltage generation.

To ascertain our understanding of the driving mechanism of metamaterial thermoelectric conversion, we calculated the local heat temperatures generated on the MA using COMSOL Multiphysics software. The details of the calculation were described in the previous literature [24]. The calculated local heat temperature of 0.16 K was obtained on the MA area, which is of a comparable level to the additional temperature gradient estimated by the output voltage generated at the environment temperature of 364 K. In contrast, heat generation corresponding to the temperature of 0.02 K was observed on the surface of the control electrode.

Table 6.1 presents the calculated and measured local temperatures generated at each element of the MA thermoelectric device. We considered that the local temperatures generated on the $Bi_{0.3}Sb_{1.7}Te_3$ element would not affect the thermoelectric performance of the MA device for the following reasons. The $Bi_{0.3}Sb_{1.7}Te_3$ thermoelectric element has an isotropic shape, meaning that the thermal radiation absorption of $Bi_{0.3}Sb_{1.7}Te_3$ does not induce temperature gradients across the element owing to uniform local heat generation across the element. On the other hand, the local heat generated on the MA electrode creates an anisotropic thermal distribution across the thermoelectric device. This means that the local heat generated on the MA electrode is the key to determining the output voltage across the device, regardless of the amount of temperature generated on the $Bi_{0.3}Sb_{1.7}Te_3$ thermoelectric element. As a proof of this concept, we carried out a thermoelectric simulation.

TABLE 6.1   Calculated and Measured Local Temperatures at Each Element Consisting of the MA Thermoelectric Device

Element	Calculated Local Temperature (K)	Measured Temperature (K)
MA	0.16	0.14[a]
MA electrode (except MA area, with 60 nm-thick-CaF_2 and 150 nm-thick-Ag layers), and Control electrode (with 60 nm-thick-CaF_2 and 150 nm-thick-Ag layers)	0.02	–
Bare copper (Rear side of an electrode)	negligible	–
$Bi_{0.3}Sb_{1.7}Te_3$	0.34	0.13[b]

[a] Estimated by the measured Seebeck coefficient.
[b] Measured temperature using thermistors.

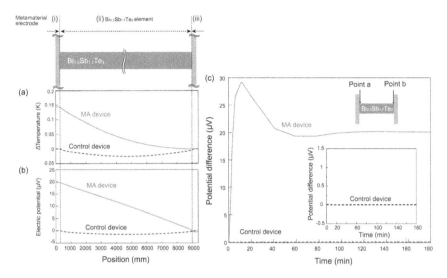

FIGURE 6.5   (a) Temperature difference and (b) electric potential distributions of the MA device (solid line) and the control device (dashed line) calculated at 180 min. The area (i), (ii), and (iii) corresponds to the MA electrode, the $Bi_{0.3}Sb_{1.7}Te_3$ element, and the control electrode, respectively. (c) Calculated time dependence of the potential differences between points a and b (the upper inset) of the MA device (solid line) and a control device (dashed line). Point a is close to the MA electrode for the MA device. The right lower inset is an enlarged potential difference graph of the control device.

Figure 6.5(a) and (b) present simulated temperature and electric potential distributions across the MA thermoelectric device and control device. Position X = 0 mm corresponds to the MA electrode surface of the MA device or a left-side electrode surface of the control electrode. The thermal distribution of the MA device in Figure 6.5(a) showed a temperature difference of 0.14 K between the electrode surfaces, which shows a good agreement with the temperature difference obtained by experiments, 0.14 K. In contrast, the control device indicated that a temperature gradient has not been induced.

The calculated electric potential distributions across the MA thermoelectric device shown in Figure 6.5(b) showed a potential difference of 19.6 µV, which is comparable to the output voltage experimentally obtained at an environmental temperature of 364 K. Meanwhile, the control device did not show a potential difference between the electrodes.

Figure 6.5(c) presents the time dependence of the potential differences between points a and b (Figure 6.5(c), upper inset) of the MA thermoelectric device and the control device. A potential difference of 20 µV in the MA thermoelectric was observed at 180 min. In contrast, the potential difference observed for the control device is almost negligible, approximately 2.0 nV. Thermoelectric simulation supports our hypothesis that the MA triggers the generation of a thermal gradient across the thermoelectric device placed in an environment with uniform thermal radiation.

6.4 NECESSITY OF METAMATERIAL ABSORBER TO DRIVE THERMOELECTRIC CONVERSION IN AN ENVIRONMENT WITH UNIFORM THERMAL RADIATION

The critical factors for the metamaterial thermoelectric conversion are the gathering of thermal energy from the surrounding media and the conduction of plasmonic local heat to a thermoelectric element. These factors can be achieved by the optical thickness and thin geometrical structure of the metamaterial, facilitate the effective conduction of the local heat to the Cu electrode. This indicates that the MA is a promising candidate for facilitating thermoelectric conversion in an environment with uniform thermal radiation. However, in the above investigation, we cannot answer two questions at this moment.

1. How important is the thin geometrical structure of an absorber that drives the metamaterial thermoelectric conversion?

2. Can any other absorber drive thermoelectric conversion in an environment with uniform thermal radiation?

To examine whether other effective absorbers can drive thermoelectric conversion in an environment with a uniform thermal radiation and clarify how metamaterial features contribute to thermoelectric conversion, we compared the thermoelectric performance of devices loaded with the MA and a carbon black (CB) layer, which is a more effective IR absorber. Here, MA is a material that satisfies the optical thickness and thin geometrical structure requirements. In contrast, CB only meets the requirement for optical thickness, but not thin geometrical structure. By comparing the thermoelectric performance of the devices loaded with these materials, we can discuss the significance of the metamaterial characteristics in the enhancement of thermoelectric devices driven in an environment with uniform thermal radiation.

For comparison, we prepared a CB device by replacing the MA electrode of the MA device with a CB-coated electrode. To create a CB-coated Cu electrode, we coated a CB layer using a CB spray (TA410KS, Ichinen

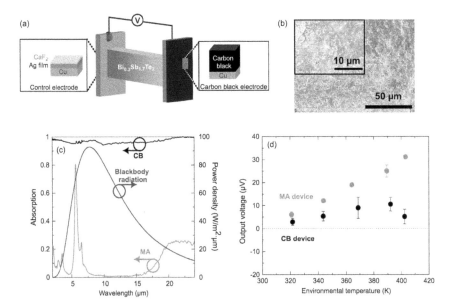

FIGURE 6.6 (a) Schematic of the $Bi_{0.3}Sb_{1.7}Te_3$ thermoelectric device loaded with the CB electrode on the right side of the thermoelectric element and (b) top-view SEM images of the CB electrode. (c) Absorption spectra of the MA and 100-μm-thick CB electrodes. A blackbody radiation spectrum calculated at 364 K was shown for a comparison. (d) Dependence of output voltages generated on the MA device and a 100 μm-CB device on the measured environmental temperature.

Tasco Co., Ltd.) with an emissivity of 0.94 on a Cu electrode with an arbitrary thickness (Figure 6.6(a), right side). The measured thickness of the CB layer is 103 ± 1 μm. Figure 6.6(b) shows top-view images of the CB surface. The CB-coated area corresponds to the Cu electrode area because we coated CB on the whole Cu electrode (4×6 mm²), which is considerably larger than the MA area (2.1×2.1 mm²).

Figure 6.6(c) compares the absorption spectra of the MA and CB electrodes. The MA showed an absorption peak at 6.7 μm, whereas the control electrode indicated by a dashed line did not show significant IR absorption in this wavelength region. The CB layer showed a broader and stronger absorption than the MA in the IR range. The blackbody radiation spectrum calculated at 364 K was showed for a comparison, which indicates that the CB layer absorbs more thermal radiation than the MA.

Figure 6.6(d) shows the dependence of the output voltages of the MA device on the measured environmental temperature. The MA device exhibited an output voltage of 19.0 ± 0.1 μV (measured number $n = 7$) at an environmental temperature of 364 K. Based on the measured Seebeck coefficient of the $Bi_{0.3}Sb_{1.7}Te_3$ element (-140 μV/K), the MA induced a temperature difference of 0.14 K between the ends of the thermoelectric element. In contrast, the 100-μm-thick CB device generated an output voltage of 9.00 ± 4.6 μV at the environmental temperature of 364 K, which was smaller than that generated on the MA device. We measured output voltages generated on these two devices at different environmental temperatures ranging from 320 K to 403 K. Consequently, the MA device exhibited higher output voltages at every environmental temperature than those generated on the 100-μm-thick CB device (Figure 6.6(d)). The output voltages generated on the MA device increased linearly with an increase in environmental temperature. In contrast, the output voltages generated on CB devices showed no temperature dependence.

Based on these characteristics, we hypothesized the mechanisms of the thermoelectric conversion driven by the MA and CB electrodes. The MA electrode absorbs less thermal radiation than the CB electrode because of its narrow band absorption; however, the MA has an ultrathin structure with high thermal conductivity [25, 26], which leads to effective conductive local heat propagation for heating the Cu electrode. In contrast, CB has a higher heat power because of its broader absorption in the IR region. However, sufficient local heat is not conducted to the Cu electrode because of the lower thermal conductivity [27] and larger film thickness of the CB layer. Therefore, it is expected that the amount of thermal radiation emitted from the CB surface may increase because of local heat retention at its surface.

These results indicate that metamaterial with an optical thickness and high thermal conductance property are essential for driving thermoelectric conversion in terms of thermal radiation absorption and effective conductive local heat propagation.

6.5 SUMMARY

In this research, we induced a thermal gradient across a thermoelectric element through the absorptivity control of both electrodes. The unbalanced absorptivity difference between the electrodes produced by metamaterials can lead to a thermal distribution across a certain object or space. Metamaterial thermoelectric conversion can enhance the thermoelectric performance by increasing the thermal gradient across a thermoelectric device by up to 14% in an environment of 364 K, meaning that metamaterials can enhance the performance of conventional thermo-electric devices installed in a conventional site where a thermal gradient is maintained. It should be noted that the metamaterial thermoelectric conversion cannot drive under uniform-temperature environments with-out thermal radiation, such as space. Furthermore, we concluded that the optical thickness and high thermal conductance property of the MA are critical for driving thermoelectric conversion in an environment with a uniform thermal radiation. Moreover, our findings revealed that the ther-moelectric performance of the MA thermoelectric surpassed that of the CB device. This thermal engineering technique can be applied to control thermal distributions in microfluidics, electric circuits, and containers that are filled with uniform-temperature media [28–31].

REFERENCES

[1] Fitriani et al., "A review on nanostructures of high-temperature thermoelec-tric materials for waste heat recovery." *Renew. Sust. Energ. Rev.* 64 (2016): 635–59.
[2] Jian He et al., "Advances in thermoelectric materials research: Looking back and moving forward." *Science* 357 (2017): eaak9997.
[3] Mohamed Amine Zoui et al., "A Review on Thermoelectric Generators: Progress and Applications." *Energies* 13 (2020): 3606.
[4] Xiao-Lei Shi et al., "Advanced Thermoelectric Design: From Materials and Structures to Devices." *Chem. Rev.* 120 (2020): 7399–515.
[5] Md. Nazibul Hasan et al., "Inorganic thermoelectric materials: A review." *Int. J. Energy Res.* 44 (2020): 6170–222.
[6] Matteo Massetti et al., "Unconventional thermoelectric materials for energy harvesting and sensing applications." *Chem. Rev.* 121 (2021): 12465–547.
[7] Canlin Ou et al., "Fully printed organic–inorganic nanocomposites for flexible thermoelectric applications." *ACS Appl. Mater. Interfaces* 10 (2018): 19580–87.
[8] Byeongmoon Lee et al., "High-performance compliant thermoelectric generators with magnetically self-assembled soft heat conductors for self-powered wearable electronics." *Nat. Commun.* 11 (2020): 5948.
[9] Shohei Katsumata et al., "Metamaterial perfect absorber simulations for intensifying the thermal gradient across a thermoelectric device." *Opt. Express* 29 (2021): 16396–405.
[10] Claire M. Watts et al., "Metamaterial Electromagnetic Wave Absorbers." *Adv. Mater.* 24 (2012): OP98–OP120.
[11] Nina Meinzer et al., "Plasmonic meta-atoms and metasurfaces." *Nat. Photon.* 8 (2014): 889–98.
[12] Alexander V. Kildishev et al., "Planar photonics with metasurfaces." *Science* 339 (2013): 1232009.
[13] N. I. Landy et al., "Perfect metamaterial absorber." *Phys. Rev. Lett.* 100 (2008): 207402.

[14] Shohei Katsumata et al., "Effect of metamaterial perfect absorber on device performance of PCPDTBT:PC71BM solar cell." *Physica Status Solidi (a)* 217 (2020): 1900910.

[15] Guillaume Baffou et al., "Applications and challenges of thermoplasmonics." *Nat. Mater.* 19 (2020): 946–58.

[16] Mark L. Brongersma et al., "Plasmon-induced hot carrier science and technology." *Nat. Nanotechnol.* 10 (2015): 25–34.

[17] Gregory V. Hartland et al., "What's so hot about electrons in metal nanoparticles?" *ACS Energy Lett.* 2 (2017): 1641–53.

[18] Guillaume Baffou et al., "Thermo-plasmonics: using metallic nanostructures as nano-sources of heat." *Laser Photonics Rev.* 7 (2013): 171–87.

[19] Kelly W. Mauser et al., "Resonant thermoelectric nanophotonics." *Nat. Nanotechnol.* 12 (2017): 770–77.

[20] Ali Sobhani et al., "Narrowband photodetection in the near-infrared with a plasmon-induced hot electron device." *Nat. Commun.* 4 (2013): 1643.

[21] Haibin Tang et al., "Plasmonic hot electrons for sensing, photodetection, and solar energy applications: A perspective." *J. Chem. Phys.* 152 (2020): 220901.

[22] Christian Kuppe et al., ""Hot" in plasmonics: Temperature-related concepts and applications of metal nanostructures." *Adv. Opt. Mater.* 8 (2020): 1901166.

[23] Satoshi Ishii et al., "Titanium nitride nanoparticles as plasmonic solar heat transducers." *J. Phys. Chem. C* 120 (2016): 2343–48.

[24] Takuya Asakura et al. Metamaterial thermoelectric conversion. arXiv:2204. 13235 (2022). https://ui.adsabs.harvard.edu/abs/2022arXiv220413235A

[25] N. Heuck et al., "Analysis and modeling of thermomechanically improved Silver-Sintered die-attach layers modified by additives." *IEEE Trans. Compon. Packag. Manuf. Technol.* 1 (2011): 1846–55.

[26] Pushparajah Rajaguru et al., "Sintered silver finite element modelling and reliability based design optimisation in power electronic module." *Microelectron. Reliab.* 55 (2015): 919–30.

[27] Zhidong Han et al., "Thermal conductivity of carbon nanotubes and their polymer nanocomposites: A review." *Prog. Polym. Sci.* 36 (2011): 914–44.

[28] Ying Li et al., "Transforming heat transfer with thermal metamaterials and devices." *Nat. Rev. Mater.* 6 (2021): 488–507.

[29] B. Ciraulo et al., "Long-range optofluidic control with plasmon heating." *Nat. Commun.* 12 (2021): 2001.

[30] Wei Li et al., "Nanophotonic control of thermal radiation for energy applications [Invited]." *Opt. Express* 26 (2018): 15995–6021.

[31] Jun Wang et al., "Thermal metamaterial: Fundamental, application, and outlook." *iScience* 23 (2020): 101637.

CHAPTER 7

Thermal Emission Control Based on Photonic Nanostructures Toward High-Power High-Efficiency Thermophotovoltaic Power Generation

Takuya Inoue, Takashi Asano, and Susumu Noda

Kyoto University, Kyoto, Japan

7.1 INTRODUCTION

Thermal emitters generally exhibit a broad spectrum covering a very wide range of frequencies from ultra-violet to THz waves and are used for various applications such as thermophotovoltaic (TPV) power generation [1, 2], infrared spectroscopy [3], and thermal imaging [4]. Among these applications, TPV systems, which convert heat into electricity by irradiating PV cells with thermal emission from heated objects at high temperatures, are attracting increasing attentions as a promising candidate for next-generation power generation systems for sustainable society. TPV systems can use various heat sources, such as solar power and wasted heat in industries. In addition, they are potentially capable of providing high output power densities and high conversion efficiencies in a compact system. However, thermal emission spectra from typical emitters are too broadband compared to the spectral range where photovoltaic (PV) cells can efficiently convert light into electricity; the thermal emission at photon energies below the bandgap energy of the PV cell cannot generate electron-hole pairs inside the PV cell, while the excess energy of the photon energies above the bandgap energy will be converted into phonons (heat) inside the PV cells. To increase the power density and the power conversion efficiency of the TPV systems, it is necessary to concentrate the thermal radiation spectrum in the desired wavelength range. More specifically, it is important to enhance the thermal emission at photon energies just above the bandgap energy of the PV cell, while suppressing those at energies below the bandgap as shown in Figure 7.1 [5]. With such a spectral control of thermal emission, the ratio of the thermal emission power that can contribute to the electrical power generation can be dramatically increased.

One powerful strategy to equivalently realize the above-mentioned spectral control is the use of reflectors or spectral filters that can reflect the unnecessary spectral components into the emitter for energy recycling. For example, by introducing a gold reflector at the bottom of an InGaAs PV cell (bandgap wavelength of 1.68 μm) and inserting an air gap between the reflector and the PV cell, an average reflectivity of >95% for below-bandgap energies has been demonstrated [6]. Furthermore, a tandem PV cell (bandgap wavelengths of 0.89 and 1.03 μm) equipped with a gold back-reflector (without an air gap) has also achieved an average reflectivity

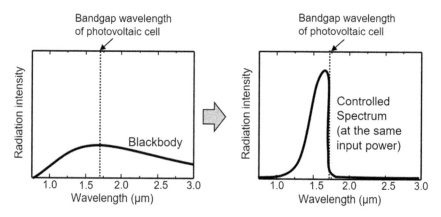

FIGURE 7.1 Schematic of thermal emission control for thermophotovoltaic applications.

of 93% [7]. However, in this strategy, the design of the entire TPV system becomes complicated because it is necessary to return all the light reflected by the PV cell (or spectral filters) to the thermal radiation source without dissipating it to the outside of the system. Therefore, most of the previous research based on this strategy did not report the efficiency of the entire TPV system but focused on the evaluation of the efficiency of the PV cell. On the other hand, if the radiation spectrum of the thermal radiation source itself can be controlled to a narrowband spectrum as shown in Figure 7.1, it is expected to realize a highly efficient TPV system with a simpler and more compact system. In this chapter, we discuss in detail about the latter strategy.

The principle of thermal emission control is based on Kirchhoff's law of thermal radiation [8], which states that the emissivity of an emitter is equal to the absorptivity of the same emitter for a given frequency, direction, and polarization. Therefore, to realize a narrowband thermal emission spectrum, we have to increase the absorptivity of the emitter only at a target wavelength, while suppressing the background absorptivity at all other wavelengths. One approach for obtaining narrowband absorptivity is the use of materials that inherently cause strong absorption at fixed wavelengths such as rare earth oxides [9]. However, this approach does not enable the flexible control of an emission wavelength or an emission bandwidth. On the other hand, absorptivity is also dependent on the magnitude of interaction between light and materials inside the emitter; one can increase absorptivity at a particular wavelength by using an optical resonance which enhances the light-matter interaction at that wavelength. Based on this approach, various types of wavelength-scale optical nanostructures such as photonic crystals or metamaterials have been utilized for thermal emission control in the past decades. Specifically, nanostructures based on refractory metals have been most intensively investigated in TPV applications [10] because refractory metals are stable even at high temperatures and show larger absorption in the near-infrared than at the longer wavelengths, which are potentially suitable for near-infrared selective thermal emitters in TPV applications. It should be noted, however, that the suppression of background absorptivity (emissivity) in those metallic nanostructures is not perfect due to the strong free carrier absorption in the nanostructured metals, especially at high temperatures, which reduces the conversion efficiency of the TPV systems.

To enable the full control of a thermal emission spectrum, thermal emission control based on simultaneous control of material absorption and photonic resonances was proposed [5, 11]. For example, in the mid-infrared region, which is important for infrared spectroscopy and environmental sensing applications, thermal emitters that utilize the intersubband transition of quantum wells and the resonance mode of photonic crystals were developed [5, 12, 13]. Based on this approach, an ultra-narrowband thermal emission spectrum whose linewidth is less than 1/100 of that of the blackbody spectrum was experimentally demonstrated [12, 13], and high-speed electrical control of thermal emission intensity and thermal emission wavelengths were also realized [14, 15]. In addition, by extending this method to the near-infrared region, near-infrared narrowband thermal emitters that can be applied to TPV applications were also developed [16–18].

On the other hand, the thermal radiation intensity that can be extracted into free space is always limited by the blackbody radiation intensity determined by Planck's law at the same temperature (blackbody limit). Since all thermal emitters have an upper temperature limit that can be heated, the generated power density obtained in TPV systems is also limited by the above-mentioned blackbody limit. In order to overcome this issue, the use of near-field thermal radiation transfer in the TPV system (near-field TPV) has been attracting increasing attention [19, 20]. In the near-field TPV, an emitter and a PV cell are separated by a gap smaller than the characteristic wavelength of thermal emission, where not only the modes propagating in free space but also the modes confined inside the emitter and PV cell (frustrated modes) or localized at the surface can contribute to the energy transfer via evanescent coupling. Furthermore, when the use of near-field thermal radiation transfer is combined with the above-mentioned strategies of the thermal radiation spectrum control, one can realize a thermal radiation spectrum that exceeds the blackbody limit only at the desired wavelength, which enables the effective use of thermal energy in various applications, including high-power high-efficiency TPV systems.

Here, we describe the thermal emission control based on photonic nanostructures toward high-power high-efficiency TPV power generation. In Section 7.2, near-infrared narrowband thermal emitters based on a silicon photonic crystal is developed, and the results of demonstrating a highly efficient TPV system using the developed emitters are introduced. In Section 7.3, to further increase the output power density of the TPV systems, a TPV device using near-field thermal radiation transfer in a nano-gap is developed, and the photocurrent generation exceeding the blackbody limit at the same temperature is demonstrated.

7.2 NEAR-INFRARED NARROWBAND THERMAL EMITTERS BASED ON Si PHOTONIC CRYSTALS FOR FAR-FIELD TPV SYSTEMS

7.2.1 Development of Near-Infrared Narrowband Thermal Emitters Based on Si Photonic Crystals

As described in the previous section, for TPV applications, it is important to develop thermal emitters that selectively radiate in the near-infrared to visible range, which corresponds to the bandgap wavelength of typical PV cells. For this purpose, one should select emitter materials that can be used at temperatures higher than 1000°C, which is necessary for the generation of near-infrared thermal emission. In addition, the near-infrared absorption coefficient of the materials should be high enough to achieve the high absorptivity (emissivity) while the absorption coefficients at the longer wavelengths should be low to suppress the background absorptivity (emissivity). Considering these requirements, intrinsic silicon (Si) is a suitable choice for the materials of selective thermal emitters in the near-infrared range [16]. Si has a relatively high melting point (1683 K), and it shows a large absorption coefficient in the visible and near-infrared range owing to the interband transition. Figure 7.2(a) shows the calculated absorption coefficient of intrinsic Si for various temperatures [16]. Here, both interband transitions and free-carrier absorption due to thermally excited intrinsic carriers are considered. As shown in Figure 7.2(a) , the bandgap wavelength of Si redshifts as the temperature rises. The subbandgap absorption coefficient due to thermally excited intrinsic carriers increases as the temperature rises, but it is still one to two orders of magnitude smaller than that of interband absorption. Therefore, by choosing the appropriate thickness of the Si emitter, one can obtain high absorptivity (emissivity) in the near-infrared range and relatively low absorptivity (emissivity) in the longer wavelengths.

To further increase the frequency selectivity of the thermal emission, a rod-type photonic crystal resonator was introduced into the Si slab as shown in the inset of Figure 7.2(b), so that one can resonantly enhance the absorptivity in the near-infrared range. Rod-type photonic crystals can reduce the Si filling factor, which is important for the suppression of the nonresonant longer-wavelength emission from the intrinsic free carriers

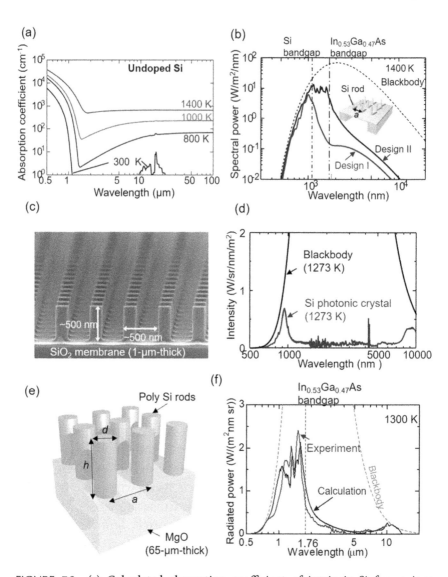

FIGURE 7.2 (a) Calculated absorption coefficient of intrinsic Si for various temperatures. (b) Calculated radiation spectra of the two designed Si rod-type photonic crystals (Design I: $a = 600$ nm, $h = 600$ nm, $r = 105$ nm, Design II: $a = 700$ nm, $h = 800$ nm, $r = 190$ nm), which were integrated over the upper hemisphere. (c) Scanning microscope image of a Si rod-type photonic-crystal thermal emitter. (d) Measured thermal radiation spectrum of the emitter shown in (c) and the blackbody spectrum at 1273 K. (e) Schematic of a Si rod-type photonic-crystal thermal emitter with a MgO substrate. (f) Measured and calculated thermal radiation spectra of the emitter shown in (e).

generated at elevated temperatures. Here, the lattice constant (a), rod radius (r), and rod height (h) of the photonic crystal were appropriately designed so that the structure supports resonant modes in the visible-to-near-infrared range. Figure 7.2(b) shows the calculated radiation spectra of the two designed Si rod-type photonic crystals (Design I: $a = 600$ nm, $h = 600$ nm, $r = 105$ nm, Design II: $a = 700$ nm, $h = 800$ nm, $r = 190$ nm), which were integrated over the upper hemisphere. As shown in Figure 7.2(b), the designed Si rod-type photonic crystals show frequency-selective thermal emission in the near-infrared range, while the background emission intensity is much suppressed. More specifically, Design I exhibits a frequency-selective thermal emission spectrum around the bandgap of Si PV cells (1100 nm), while Design II, which contains larger silicon rods than Device I, exhibits a thermal emission spectrum concentrated above the bandgap of $In_{0.53}Ga_{0.47}As$ PV cell (~1700 nm).

Figure 7.2(c) shows a scanning electron microscope image of the fabricated Si rod-type photonic crystal [16]. Here, both the lattice constant and rod height were set to 500 nm so that the wavelength of the resonance peak was located around 1 μm. A 1-μm-thick SiO_2 membrane was used to support the array of Si rods in the fabricated structure. The fabricated emitter was heated by the external ceramic heater in an Ar-filled chamber, and the thermal emission spectrum of the emitter in the vertical direction was measured using a grating-type spectrometer equipped with Si and InGaAs detector arrays and a Fourier-transform infrared (FTIR) spectrometer equipped with a HgCdTe detector. Figure 7.2(d) shows the measured thermal emission spectra of the fabricated Si photonic crystal and a reference blackbody emitter at a temperature of 1273K. The Si photonic crystal shows a narrowband thermal emission peak with a near-unity emissivity at a wavelength of around 1 μm, while it shows relatively suppressed thermal emission in a wavelength range from 1 μm to 6 μm. This result clearly shows the advantage of the thermal emission control based on intrinsic Si and photonic crystals. It should be noted that the thermal emission at wavelengths longer than 6 μm is caused by the emission from the SiO_2 membrane. Therefore, it is desirable to further reduce the thickness of SiO_2, although it is technically difficult due to the mechanical vulnerability of the emitter.

In order to simultaneously ensure the mechanical robustness and optical transparency at longer wavelengths, thermal emitters based on Si rod-type photonic crystals with a MgO substrate were also developed [17, 18]. The schematic of the developed emitter was shown in Figure 7.2(e). MgO has a sufficiently high melting temperatures (3125 K), and its refractive index is much lower than Si so that it can support optical modes confined in Si rods. Moreover, MgO has smaller absorption coefficients than SiO_2 in the wavelength range of 5~20 μm, and thus we can increase the thickness of the MgO underneath the Si rod-type photonic crystals and improve the mechanical robustness of the emitter. In the fabrication, polycrystalline-Si was deposited on the MgO via low-pressure chemical vapor deposition (LP-CVD). A thin HfO_2 layer (30 nm) was deposited on the MgO before the deposition of polycrystalline-Si so that MgO and oxidized Si (SiO_2) do not react each other. Here, the lattice constant, rod height, and rod radius were adjusted so that the resonant wavelength is located around 1.8 μm, which coincides with the bandgap wavelength of $In_{0.53}Ga_{0.47}As$ PV cell, to enable the demonstration of a high-efficiency TPV system (discussed later). After fabricating rod-type photonic crystals, the MgO substrate was thinned down to the thickness of 65 μm so that the absorptivity (emissivity) of the substrate can be suppressed while the mechanical robustness is maintained. Figure 7.2(f) shows the measured thermal emission spectrum in the vertical direction when the emitter was heated at 1300 K [18]. In the same figure, the blackbody spectrum at 1300 K and the calculated thermal emission spectrum are also plotted. As shown in Figure 7.2(f), the fabricated structure selectively shows strong thermal emission at wavelengths shorter than the bandgap wavelength of $In_{0.53}Ga_{0.47}As$, which agrees well with the calculated spectrum.

7.2.2 Demonstration of High-Efficiency TPV System

Using the frequency-selective near-infrared Si thermal emitters (shown in Figure 7.2(e)) developed in the previous subsection, a prototype TPV system was experimentally constructed [18]. In this system, an $In_{0.53}Ga_{0.47}As$ PV cell with a bandgap wavelength of 1.76 μm was employed so that the emission from the Si rod-type photonic-crystal thermal emitter could be efficiently converted to electrical power. Figure 7.3(a) shows the schematic of the constructed TPV system, where a Si photonic-crystal thermal emitter is surrounded by two InGaAs PV cells. The size of the Si emitter and the InGaAs PV cell is 4.8 mm × 4.8 mm and 6.4 mm × 5.9 mm, respectively, and the separation between them is 200~300 μm. The entire system was installed into a vacuum chamber with a pressure of ~3 × 10^{-2} Pa to suppress the thermal convection loss. In the prototype system, the thermal emitter is heated via Joule heating by current injection so that the input heating power of the emitter can be precisely evaluated; by doing so, the

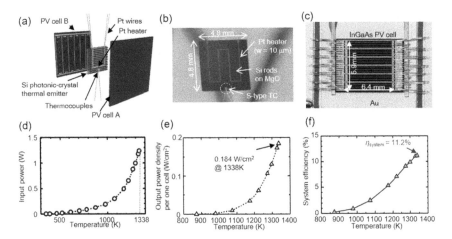

FIGURE 7.3 (a) Schematic of a far-field TPV system composed of a Si photonic-crystal thermal emitter and two InGaAs PV cells. (b)(c) Images of the fabricated Si photonic-crystal emitter and an InGaAs PV cell. (d) Measured emitter temperature as a function of input power. (e) Measured output power density per one PV cell as a function of the emitter temperature. (f) Measured system efficiency as a function of the emitter temperature.

system conversion efficiency (which is defined as the ratio of the output electrical power of the PV cell to the input heating power) can be experimentally evaluated. Figure 7.3(b) and (c) show the microscope images of the fabricated Si thermal emitter and InGaAs PV cells. As shown in Figure 7.3(b), the thermal emitter was supported by thin Pt wires whose diameter is 25 µm so that current injection was enabled while thermal conduction loss through the wire was suppressed as much as possible. In addition, by attaching handmade S-type thermocouples composed of $Pt/Pt_{0.9}Rh_{0.1}$ to the emitter, the temperature of the emitter was measured in real time. The InGaAs PV cell shown in Figure 7.3(c) was attached to the copper heatsink, whose temperature was kept at 10°C for thermal management. The measured series and shunt resistances of the PV cell were 0.01 Ωcm^2 and 1 $k\Omega cm^2$, respectively, and the measured external quantum efficiency at the wavelengths of 1.0~1.7 µm was larger than 90%.

Figure 7.3(d) shows the measured emitter temperatures as a function of the input heating power of the emitter. The emitter temperature monotonically increased as the input power increased, reaching 1338 K at an input power of 1.24 W, which was the maximum applied power in the experiment. The measured output power density per one PV cell as a function of the emitter temperature was plotted in Figure 7.3(e). The power density exponentially increased as the temperature increased, and a power density of 0.184 W/cm² was obtained at the maximum emitter temperature of 1338 K. This value is an order of magnitude higher than that of a typical PV cell in a usual solar power generation system (~0.02 W/cm²), which shows one of the advantages of TPV systems. It should be noted that the size of each PV cell (0.378 cm²) is larger than that of the emitter (0.230 cm²) in the above prototype system to maintain a large view factor. By decreasing the distance between the emitter and the PV cell, one can reduce the size of the PV cell while maintaining the same view factor, which can further increase the output power density of the PV cell.

Figure 7.3(f) shows the calculated system conversion efficiency of the constructed TPV system, which was obtained by dividing the output power of the two PV cells by the input heating power. The efficiency exceeded 10% at the emitter temperature of 1300 K, and a measured efficiency of 11.2% was obtained at the maximum input power. This efficiency was 1.65 times higher than the maximum value among the system efficiencies that had been previously reported in the experiments (6.8%) [2].

In the above prototype system, thermal radiation loss outside the view factor of the PV cells (such as the radiation from the edges of the thermal emitter) and thermal conduction loss from the supporting thin Pt wires exist, which are equivalent to 15~20% of the total input power. By scaling up the TPV system, these losses can be suppressed and the system efficiency can be further increased. It should also be noted that the PV cell used in the above experiment has a grid of front-side electrodes, which also blocks some of the reflected light at the bottom electrode. By employing a back-surface contact scheme with near-unity reflectance [6], the ideal system efficiency for the above system can become ~20%. In addition, by employing a scheme of the near-field thermal radiation transfer explained in the next subsection, both the output power density and the conversion efficiency can be further increased.

7.3 DEVELOPMENT OF ONE-CHIP NEAR-FIELD TPV DEVICE INTEGRATING Si THERMAL EMITTERS AND INGAAS PV CELLS

7.3.1 Principle of Near-Field Thermal Radiation Transfer and Previous Demonstrations

As described in the introduction, the thermal radiation intensity in free space is usually limited by the blackbody intensity $I_{\mathrm{BB}} = \dfrac{c_0}{4\pi}\hbar\omega D(\omega)n_{\mathrm{BE}}(\omega,T)$, where c_0 is the speed of light in vacuum, $\hbar\omega$ is the photon energy, $D(\omega) = \dfrac{\omega^2}{\pi^3 c_0^3}$ is the photonic density of states in vacuum and n_{BE} $(\omega, T) = \left\{ \exp\left(\dfrac{\hbar\omega}{k_B T}\right) - 1 \right\}^{-1}$ is the Bose–Einstein distribution. When the refractive index n_r of the space is larger than unity, this limit can be overcome because $D(\omega)$ is proportional to n_r^3 while the speed of light is proportional to $1/n_r$, resulting in n_r^2 times enhancement of thermal radiation power. Therefore, when two high-n_r objects (e.g., thermal emitter and PV cell) are placed closely enough to enable evanescent coupling of light, the thermal radiation transfer between them can be greatly enhanced. In addition, further enhancement ($>n_r^2$) could be achieved when we consider the contribution of the surface modes, which cannot propagate inside the objects but are localized at the surface of them. These phenomena are called near-field thermal radiation transfer (NFTRT), and its application to TPV systems is attracting attention these days.

The experimental demonstration of near-field thermal radiation transfer was technically difficult because a subwavelength gap should be maintained between two objects with different temperatures. However, recent progress in nanotechnology has led to a series of experimental demonstrations of near-field TPVs. The first experimental investigation of a near-field TPV principle was reported in 2001 [21], and several quantitative characterizations of near-field TPV systems have been reported in recent years [22–25]. The prototype near-field TPV system developed in 2018 using a piezo-controlled setup achieved a 60-nm gap between a Si-bulk thermal emitter (~660 K) and a mid-infrared photodetector (300 K) [22]. Based on the similar approach, a power density of ~5 kW/m² (= 0.5 W/cm²) with an estimated sub-system efficiency of 6.8% has been demonstrated with a thin-film InGaAs PV cell [24]. Another proof-of-concept study reported a power density of ~7.5 kW/m² (= 0.75 W/cm²) using an indium antimonide PV cell cooled to 77 K [25]. Furthermore, a precise control of the position of a thermal emitter that is within a subwavelength distance from a room-temperature germanium photodetector has been demonstrated by using a nano-electromechanical system (NEMS) [23]. It should be noted, however, that these demonstrations were performed in relatively small (<150 μm) systems and they also required external actuations such as a piezo-controlled stage and NEMS, which were not suitable for practical applications.

7.3.2 Design of One-Chip Near-Field TPV Device

To overcome the above issues, one-chip near-field TPV devices that integrated both a thermal emitter and a PV cell with a subwavelength gap were developed [26, 27]. Figure 7.4(a) shows the schematic of the developed one-chip near-field TPV device, where a 20-μm-thick undoped Si thermal emitter and a 2.9-μm-thick InGaAs PV cell were integrated with a 50-μm-thick intermediate Si substrate. The sub-wavelength gap (d) was introduced between the Si thermal emitter and Si intermediate substrate to enable evanescent coupling of light between them. The purpose of the introduction of the intermediate substrate is the suppression of the far-infrared heat transfer mediated by surface modes at the surface of the PV cell, which always exists in conventional near-field TPV systems where the emitter is directly placed close to the PV cell. These surface modes are supported by transverse optical (TO) phonons and high-density free carriers in the contact layers of the PV cells, whose permittivity in the far-infrared range is negative. Such far-infrared heat transfer can be suppressed by

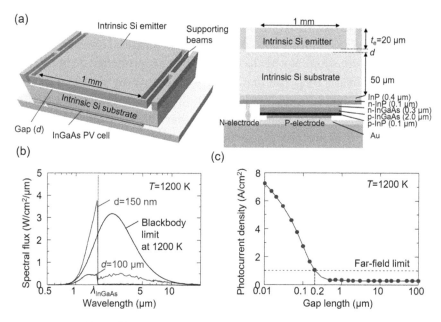

FIGURE 7.4 (a) Bird's eye view and cross section of a one-chip near-field TPV device integrating a Si thermal emitter and an InGaAs PV cell. (b) Calculated thermal radiation transfer spectrum to the InGaAs PV cell for near-field (d = 150 nm) and far-field (d = 100 μm) devices at 1200 K. The blackbody spectrum at 1200 K is also shown. (c) Calculated photocurrent density of the near-field TPV device as a function of gap length.

attaching an appropriate intermediate substrate at the top of the PV cell [28]. In this system, the enhancement of the thermal radiation transfer in the near-infrared, which contributes to the photocurrent enhancement beyond the blackbody limit, can be still achieved owing to the extraction of the frustrated modes inside the emitter into the intermediate substrate via evanescent coupling. Here, an undoped Si substrate is chosen as a material of the intermediate substrate because it is transparent at the wavelengths longer than the bandgap wavelength (1.1 µm at 300 K), and it does not support any surface modes, owing to its positive permittivity at all wavelengths. In addition, the refractive index n_r of Si in the near-infrared range (~3.5) is as high as that of InGaAs, which enables an order of magnitude enhancement of the near-field thermal radiation transfer (owing to the n_r^2 times enhancement discussed above), resulting in photocurrent generation overcoming the far-field blackbody limit. It should also be noted that the 50-µm-thick intermediate substrate is also helpful for increasing the mechanical strength of the entire device.

Unlike the Si photonic-crystal thermal emitters introduced in the previous section, the above near-filed TPV device just employs a planar Si emitter. Nevertheless, effective control of the emission spectrum is possible by recycling the below-bandgap thermal emission components from the emitter using the bottom reflector (electrode) of the PV cell, where the thicknesses of the doped layers of the InGaAs PV cell are thinned down to a few micrometers to reduce the free carrier absorption loss in the doped layers. Therefore, this device can effectively suppress the below-bandgap thermal emission loss that does not contribute to the photocurrent generation, and thus can realize high-power high-efficiency power conversion.

Figure 7.4(b) shows the calculated thermal radiation transfer spectra from the Si emitter to the InGaAs PV cell when the gap size (d) is 100 µm (corresponding to far-field case) and 150 nm (corresponding to near-field case). The emitter temperature is fixed to 1200 K. The blackbody limit determined by Planck's law at 1200 K is also plotted in the same figure. In the near-field case, the thermal radiation intensity at wavelengths shorter than the bandgap wavelength of InGaAs (λ_{InGaAs}) is seven times larger than that in the far-field case, and it exceeds the blackbody limit. It should be noted that the thermal radiation intensity at wavelengths longer than the bandgap wavelength of InGaAs due to free carrier absorption also increases in the near-field case, but its intensity is much lower than the blackbody limit, owing to the reduction of the thickness of the PV cell as described in the previous paragraph. Figure 7.4(c) shows the calculated photocurrent density of the near-field TPV device as a function of the gap length. In this calculation, it is assumed that every photon absorbed inside the p-n junction of the InGaAs layers generates one electron-hole pair. The theoretical upper limit of the photocurrent density in the far-field case at the same emitter temperature is also plotted with a dashed line. In the proposed device, one can obtain a photocurrent density exceeding the blackbody limit by reducing the gap length d below 200 nm.

7.3.3 Fabrication and Characterization of One-Chip Near-Field TPV Device

Figure 7.5(a) shows a microscope image of the fabricated Si thermal emitter with a side length of 1 mm, which was integrated on top of the intermediate Si substrate. To suspend the millimeter-sized emitter while minimizing the tilt and the thermal conduction loss, four L-shaped supporting beams (width: 10 μm, length: 580 μm) were employed. Because the width of the supporting beams (10 μm) is smaller than the thickness (20 μm), these supporting beams can relieve the thermal stress of the emitter at high temperatures by in-plane deformation. In addition, these narrow supporting beams can minimize the thermal conduction loss from the emitter and can realize relatively uniform temperature distribution inside the emitter during the heating. According to the simulation by the finite element method, this structure can reduce the in-plane variation of the vertical displacement to < 10 nm even when the emitter is heated at 1300 K. The microscope image of the fabricated 10-μm-width supporting beam is shown in the right panel of Figure 7.5(a).

To realize the integration of the emitter and the intermediate substrate, two silicon-on-insulator (SOI) wafers were bonded with a sub-wavelength gap d. The gap length d between the emitter and the intermediate substrate was controlled by the surface etching of the intermediate substrate before chip-to-chip bonding. Here, both a near-field TPV device ($d \sim 150$ nm) and a far-field TPV device ($d \sim 2900$ nm) were fabricated to quantitatively compare the generated photocurrent density in the two devices. The chip-to-chip bonding of the two SOI substrates was performed by hydrophilic bonding and subsequent annealing. After the bonding, the intermediate Si substrate was bared by removing the unnecessary Si and SiO_2 layers. Then the bonding of the intermediate substrate and the epi-structure for the PV cell (i-InP/n-InP/n-$In_{0.53}Ga_{0.47}As$/p-$In_{0.53}Ga_{0.47}As$/p-InP/p-$In_{0.53}Ga_{0.47}$ As) was performed by oxygen plasma activation and post-annealing. After removing the InP substrate, a mesa-type PV cell structure with a side

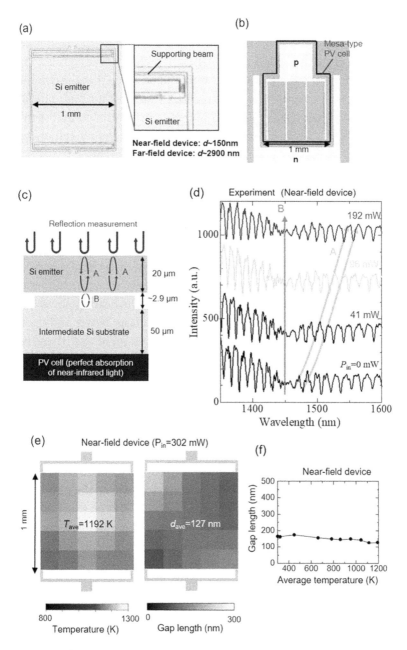

FIGURE 7.5 (a) Microscope image of a fabricated Si emitter. (b) Microscope image of a fabricated InGaAs PV cell. (c) Method for estimating the temperature of the thermal emitter and the gap between the emitter and the intermediate substrate. (d) Example of the measured reflection spectra of the Si thermal emitter with various heating powers. (e) Measured temperature distribution and gap distribution of the near-field TPV device at a heating power of 302 mW. (f) Measured gap length as a function of the emitter temperature.

length of 1 mm was formed by photolithography, metal deposition, and a lift-off process. Figure 7.5(b) shows the microscope image of the fabricated InGaAs PV cell with a side length of 1 mm, which was integrated at the bottom of the intermediate substrate. It should be noted that the comb-like electrode shown in Figure 7.5(b) is potentially suitable for high-efficiency TPV systems owing to the reduction of free carrier absorption loss in the electrodes, but it leads to a higher series resistance of the PV cell owing to the low electrical conductance of the p-type contact layers.

Before the near-field TPV experiment, the in-plane distribution of the gap length of the fabricated devices was characterized when the heating power of the Si emitter was varied. In this experiment, the heating of the emitter was performed by irradiation of the external laser light (wavelength: 532 nm, spot size 500 μm), where the device was placed in a vacuum chamber ($<1 \times 10^{-3}$ Pa) for the purpose of the proof-of-concept demonstration. In the future, the laser irradiation can be replaced with another heat source such as concentrated sunlight irradiation. The heating power of the device was calculated by considering the incident laser power and the theoretical reflectivity of Si at elevated temperatures (note that there is no transmission of the irradiated laser light owing to the sufficiently large absorption coefficient of Si). In addition, in order to estimate the gap length and the emitter temperature at each point, the spatial mapping of the reflection spectra of the emitter was measured by irradiating it with broadband infrared light. Figure 7.5(c) shows the schematic of the reflection measurement, wherein the reflection spectra at 5×5 points were measured in each device. Figure 7.5(d) shows the example of the measured spectra at various heating powers for the near-field device, where the Fabry-Perot interferences in the 20-μm-thick Si (indicated with 'A') and that in the vacuum gap (indicated with 'B') clearly appeared. Here, the former interferences redshift as the temperature rises due to the increase in the refractive index of Si, while the latter does not change because its resonant wavelength is only determined by the vacuum gap length d. Therefore, the emitter temperature and the gap length can be simultaneously estimated by comparing the measured reflection spectra and the calculated reflection spectra. It should be noted that the Fabry-Perot interferences at the gap $d <$ 500 nm do not appear in the measured wavelength range, and therefore, deeply etched sections ($d\sim2900$ nm) were partially introduced within the near-field device ($d\sim150$ nm) for the gap estimation.

Figure 7.5(e) shows the obtained temperature and gap distribution of the fabricated near-field device at the maximum heating power (302 mW).

The emitter temperature (Figure 7.5(e) shows the left panel) was highest in the center and lowest in the vicinity of the supporting beams. The maximum temperature difference within the emitter was ~100 K, which was comparable to the calculated results by the finite-element method. The obtained gap distributions (right panel) indicate that the device shows a small degree of bowing probably due to the residual stress during the fabrication process. Such bowing should be suppressed to realize a smaller gap length between the emitter and the PV cell in the future. Figure 7.5(f) shows the temperature dependence of the obtained average gap length of the fabricated near-field device. The average gap length remains almost the same for a wide range of emitter temperatures (300–1200 K), and it is smaller than 150 nm, which satisfies the condition for exceeding the blackbody limit shown in Figure 7.4(c).

7.3.4 Demonstration of Photocurrent Generation Overcoming the Blackbody Limit

Figure 7.6(a) and (b) show the measured current-voltage characteristics of the far-field TPV device (d_{ave}~2900 nm) and the near-field TPV device (d_{ave}~140 nm) for various emitter temperatures. Here, the near-field device yields much larger photocurrents than the far-field device; for example,

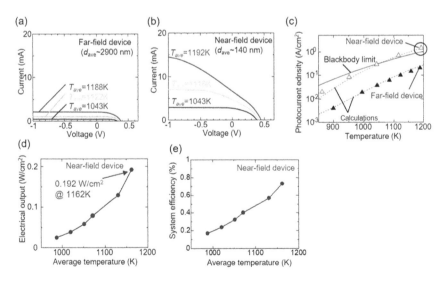

FIGURE 7.6 (a) (b) Measured current-voltage characteristics of far-field and near-field TPV devices. (c) Measured and calculated photocurrent density of far-field and near-field TPV devices. (d) Measured electrical output power density of the near-field device. (e) Measured system efficiency of the near-field device.

at the same emitter temperature of 1043 K, the short-circuit current of the near-field device (2.69 mA) is 7.3 times larger than that of the far-field device (0.37 mA). It should be noted that in the near-field device (Figure 7.6(b)), the short-circuit currents (current at a voltage of 0 V) at the emitter temperatures of 1116 K and 1192 K were less than the original photocurrents generated by the near-field thermal radiation transfer, which were measured when a sufficient reverse bias (−1 V) was applied. Since the photocurrent densities in the near-field device (>1 A/cm²) are one to two orders of magnitude higher than those of the typical solar cells (several tens of mA/cm²), even a small series resistance induces a non-negligible internal forward bias on the p-n junction of the PV cell, by which a part of the generated photocurrent is internally consumed as a forward current. To avoid this issue, the reduction of the series resistance via the modification of the electrode design is important. Figure 7.6(c) shows the measured and calculated photocurrent densities of the two devices as a function of the average emitter temperature (note that the uniform temperature distribution was assumed in the calculations). The measured photocurrent densities of the fabricated devices (triangles) agree well with the corresponding calculated ones (dashed lines), and the current densities of the near-field device is 5 to 10 times larger than those of the far-field device. Especially, the photocurrent density obtained in the near-field device is larger than the blackbody limit (solid line) when the emitter temperature is larger than 1050 K; for example, the experimentally obtained photocurrent density at 1192 K is 1.49 A/cm², which is 1.5 times larger than the blackbody limit at the same temperature. This result is the first demonstration of the photocurrent generation overcoming the far-field blackbody limit in a one-chip device without any external actuations.

Then, the electrical output power of the near-field TPV device was evaluated by changing the input heating power of the device. In this experiment, another near-field TPV device that has a uniform p-side electrode instead of a comb-like electrode shown in Figure 7.5(b) was used to decrease the series resistance of the PV cell. Figure 7.6(d) shows the measured electrical output power density of the near-field TPV device as a function of the emitter temperature. At an emitter temperature of 1162 K, the electrical power density of 0.192 W/cm² was obtained. Compared to the result of the far-field TPV experiment in the previous section (Figure 7.3(e)), where the electrical power density of 0.184 W/cm² was obtained at 1338 K, almost the same electrical power density was obtained in the near-field device even when the emitter temperature was 200 K lower than that of the far-field device. Although the above power density is smaller than the highest value ever reported in the near-field TPV experiment (0.75W/cm²) [25], the absolute value of the electrical output power (1.92 mW) is two orders of magnitude larger than those of the previously demonstrated near-field TPV devices [22–25] owing to the millimeter-sized device (1 mm²).

Finally, the actual system efficiency of the near-field TPV system was evaluated by taking the ratio between the measured input heating power of the emitter and the electrical output power. It should be noted that the previous demonstrations of near-field TPV systems [22–25] resulted in system efficiencies lower than 0.01% since they did not introduce any scheme to reduce the thermal conduction loss from the emitter. Figure 7.6(e) shows the measured system efficiency of the near-field TPV system as a function of the emitter temperature. At an emitter temperature of 1162 K, the maximum system efficiency of 0.7 % was obtained, which was one to two orders of magnitude larger than those of the previous near-field TPV systems. In the next section, the strategies to further increase the system efficiency of the near-field TPV device were discussed.

7.3.5 Strategies toward Highly Efficient Near-Field TPV Systems

Currently, the system efficiency of the one-chip near-field TPV device was mainly limited by the following three factors: (1) thermal conduction loss through the supporting beams, (2) thermal radiation loss in the direction opposite to the PV cell, and (3) free-carrier absorption loss in the bottom electrode of the PV cell. In the following section, the solutions for these three factors are discussed.

As for the thermal conduction loss (factor (1)), one should design the improved structure of the supporting beams that can further reduce the conduction loss but can still maintain the flatness of the emitter. When the width and the thickness of the supporting beams are set as w and t, respectively, the thermal conduction loss is proportional to wt while the section modulus in the vertical direction is proportional to wt^2. Therefore, one should increase the thickness t and decrease the width w of the supporting beams to simultaneously realize the small thermal conduction loss and mechanical robustness. Another approach to effectively decrease the ratio of the thermal conduction loss is the increase of the emitter temperature and the further reduction of the gap length d between the emitter and the intermediate substrate, which drastically increases the electrical output power of the PV cell.

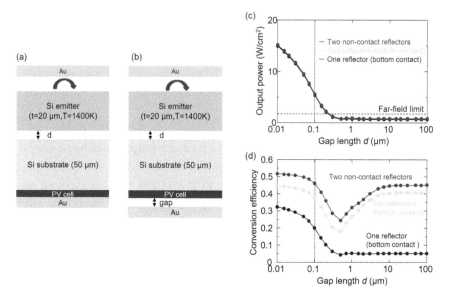

FIGURE 7.7 (a) (b) Simulation models of near-field TPV systems with top reflectors for photon recycling: (a) two reflectors in the top and bottom, where the bottom reflector is in contact with the PV cell; and (b) two reflectors in the top and bottom, where the bottom reflector is not in contact with the PV cell. (c) Calculated electrical output power density of the designed TPV systems when the emitter temperature is fixed to 1400 K. (d) Calculated conversion efficiency of the designed TPV systems when the emitter temperature is fixed to 1400 K.

To decrease the thermal radiation loss in the direction opposite to the PV cell (factor (2)), the introduction of a top reflector above the emitter for photon recycling is effective [29]. Figure 7.7(a) and (b) show the schematics of the near-field TPV device with top and bottom gold reflectors. In Figure 7.7(a), the bottom reflector is in contact with the PV cell (note that the top reflector should be separated with vacuum from the high-temperature thermal emitter), while the bottom reflector is also separated with a vacuum from the PV cell in Figure 7.7(b). From the viewpoint of the absorption loss in the bottom reflector (factor (3)), the latter configuration is more favorable because all the frustrated modes inside the PV cell are reflected by the total internal reflection at the interface of the PV cell and vacuum, and only the propagating modes inside the vacuum reach the bottom reflector [29]. Figure 7.7(c) and (d) show the calculated electrical output power density and conversion efficiency of the designed TPV systems when the emitter temperature is fixed to 1400 K. In these calculations, the series resistance of the PV cell was neglected and the dark current density of the PV cell was assumed to be the theoretical minimum value (thermodynamic limit) for the ideal case. In Figure 7.7(c), the output power density is enhanced by a factor of 3.3 and 8.8 compared to the far-field limit when the gap length is 0.1 μm and 0.01 μm, respectively. It should be noted that the introduction of the top reflectors does not contribute to the increase of the output power under the condition of the constant emitter temperature. On the other hand, in Figure 7.7(d), the introduction of the two reflectors (top and bottom) significantly increases the conversion efficiency owing to the recycling of the thermal radiation in the direction opposite to the PV cell. Specifically, by introducing the two non-contact reflectors, a conversion efficiency as high as 46.1% can be realized when the gap length is 0.1 μm. It should be noted that the conversion efficiency in the far-field case ($d > 10$ μm) also exceeds 40%, but the output power density is one order of magnitude lower than that of the near-field system as shown in Figure 7.7(c).

7.4 SUMMARY

In this chapter, recent experimental demonstrations of far-field and near-field TPV systems based on silicon photonic nanostructures have been reviewed. In the far-field experiment, silicon rod-type photonic-crystal thermal emitters that exhibit near-infrared thermal emission with suppressed background emission have been developed, and a system conversion efficiency of 11.2% has been demonstrated in the constructed prototype TPV system composed of the developed silicon emitter and InGaAs PV cells. In the near-field experiment, a one-chip near-field TPV device integrating a thin-film Si thermal emitter and InGaAs PV cell with an intermediate Si substrate has been developed, and photocurrent generation overcoming the far-field blackbody limit has been realized at an emitter temperature 1192 K. In both cases, parasitic thermal losses such as thermal radiation loss outside the view factor of the PV cells and thermal conduction loss should be suppressed as much as possible to boost the system conversion efficiency of the TPV systems. The reduction of the series resistance and dark current density of the PV cell is also important for the realization of high-power highly efficient TPV because the current density is much higher than that in conventional solar power generation. Addressing these remaining issues, compact and efficient TPV systems with higher conversion efficiencies than those of conventional solar power generation systems can be realized, and they will contribute to the realization of a sustainable society through the effective use of thermal energy.

REFERENCES

[1] Swanson, Richard M. "A proposed thermophotovoltaic solar energy conversion system." *Proc. IEEE* 67 no. 3 (1979): 446–447.

[2] Bierman, David M., Andrej Lenert, Walker R. Chan, et al. "Enhanced photovoltaic energy conversion using thermally-based spectral shaping." *Nat. Energy* 1 no. 6 (2016): 16068.

[3] Hodgkinson, Jane and Ralph P Tatam. "Optical gas sensing: A review." *Meas. Sci. Technol.* 24 no. 1 (2013): 012004.

[4] Meola, Carosena. and Giovanni M. Carlomagno. "Recent advances in the use of infrared thermography." *Meas. Sci. Technol.* 15 no. 9 (2004): R27–R58.

[5] Zoysa, Menaka D., Takashi Asano, Keita Mochizuki et al. "Conversion of broadband to narrowband thermal emission through energy recycling." *Nat. Photon.* 6 no. 8 (2012): 535–539.

[6] Fan, Dejiu, Tobias Burger, Sean McSherry et al. "Near-perfect photon utilization in an air-bridge thermophotovoltaic cell," *Nature* 586 no. 7828 (2020): 237–241.

[7] Lapotin, Alina. Kevin L. Schulte, Myles A. Steiner et al. "Thermophotovoltaic efficiency of 40%." *Nature* 604 no. 7905 (2022): 287–291.

[8] Brace, D. B. *The Laws of Radiation and Absorption: Memoirs by Prévost, Stewart, Kirchhoff, and Kirchhoff and Bunsen.* (American Book Company, 1901).

[9] Sai, Hitoshi, Hiroo Yugami, Kazuya Nakamura et al. "Selective emission of $Al_2O_3/Er_3Al_5O_{12}$ eutectic composite for thermophotovoltaic generation of electricity." *Jpn. J. Appl. Phys.* 39 no. 4R (2000): 1957–1961.

[10] Rinnerbauer, Veronika, Yi X. Yeng, Walker R. Chan, et al. "High-temperature stability and selective thermal emission of polycrystalline tantalum photonic crystals." *Opt. Express* 21 no. 9 (2013): 11482–11491.

[11] Asano, Takashi, Keita Mochizuki, Makoto Yamaguchi et al. "Spectrally selective thermal radiation based on intersubband transitions and photonic crystals." *Opt. Express* 17 no. 21 (2009): 19190–19202.

[12] Inoue, Takuya, Menaka D. Zoysa, Takashi Asano et al. "Single-peak narrowbandwidth mid-infrared thermal emitters based on quantum wells and photonic crystals." *Appl. Phys. Lett.* 102 no. 19 (2013): 191110.

[13] Inoue, Takuya, Menaka D. Zoysa, Takashi Asano et al. "High-Q mid-infrared thermal emitters operating with high power-utilization efficiency." *Opt. Express* 24 no. 13 (2016): 15101–15109.

[14] Inoue, Takuya, Menaka D. Zoysa, Takashi Asano et al. "Realization of dynamic thermal emission control." *Nat. Materials* 13 no. 10 (2014): 928–931.

[15] Inoue, Takuya, Menaka D. Zoysa, Takashi Asano et al. "On-chip integration and high-speed switching of multi-wavelength narrowband thermal emitters." *Appl. Phys. Lett.* 108 no. 9 (2016): 091101.

[16] Asano, Takashi, Masahiro Suemitsu, Kohei Hashimoto et al. "Near-infrared–to–visible highly selective thermal emitters based on an intrinsic semiconductor." *Sci. Adv.* 2 no. 12 (2016): e1600499.

[17] Suemitsu, Masahiro, Takashi Asano, Menaka D. Zoysa et al. "Wavelength-selective thermal emitters using Si-rods on MgO." *Appl. Phys. Lett.* 112 no. 1 (2018): 011103.

[18] Suemitsu, Masahiro, Takashi Asano, Takuya Inoue et al. "High-efficiency thermophotovoltaic system that employs an emitter based on a silicon rod-type photonic crystal." *ACS Photon.* 7 no. 1 (2020): 80–87.

[19] Laroche, Marine, Remi Carminati, and Jean-Jacques Greffet. "Near-field thermophotovoltaic energy conversion," *J. Appl. Phys.* 100 no. 6 (2006): 063704.

[20] Mittapally, Rohith. Ayan Majumder, Pramod Reddy et al. "Near-Field thermophotovoltaic energy conversion: progress and opportunities," *Phys. Rev. Appl.* 19 no. 3 (2023): 037002.

[21] DiMatteo, R. S., P. Greiff, S. L. Finberg et al. "Enhanced photogeneration of carriers in a semiconductor via coupling across a nonisothermal nanoscale vacuum gap." *Appl. Phys. Lett.* 79 no. 12 (2001): 1894–1896.

[22] Fiorino, Anthony, Linxiao Zhu, Dakotah Thompson et al. "Nanogap near-field thermophotovoltaics." *Nat. Nanotechnol.* 13 no. 9 (2018): 806–811.

[23] Bhatt, Gaurang R., Bo Zhao, Samantha Roberts et al. "Integrated near-field thermo-photovoltaics for heat recycling." *Nat. Commun.* 11 (2020): 2545.

[24] Mittapally, Rohith, Byungjun Lee, Linxiao Zhu et al. "Near-field thermophotovoltaics for efficient heat to electricity conversion at high power density." *Nat. Commun.* 12 (2021): 4364.

[25] Lucchesi, Christophe, Dilek Cakiroglu, Jean-Philippe Perez et al. "Near-Field thermophotovoltaic conversion with high electrical power density and cell efficiency above 14%." *Nano Lett.* 21 no. 11 (2021): 4524–4529.

[26] Inoue, Takuya, Takaaki Koyama, Dongyeon D. Kang et al. "One-chip near-field thermophotovoltaic device integrating a thin-film thermal emitter and photovoltaic cell." *Nano Lett.* 19 no. 6 (2019): 3948–3952.

[27] Inoue, Takuya, Keisuke Ikeda, Bongshik Song et al. "Integrated near-field thermophotovoltaic device overcoming blackbody limit." *ACS Photon.* 8 no. 8 (2021): 2466–2472.

[28] Inoue, Takuya, Kohei Watanabe, Takashi Asano et al. "Near-field thermophotovoltaic energy conversion using an intermediate transparent substrate." *Opt. Express* 26 no. 2 (2018): A192–A208.

[29] Inoue, Takuya, Taiju Suzuki, Keisuke Ikeda et al. "Near-field thermophotovoltaic devices with surrounding non-contact reflectors for efficient photon recycling." *Opt. Express* 29 no. 7 (2021): 11133–11143.

Passive Radiative Cooling Applications for Thermal and Electrical Energy Harvesting

Ross Y. M. Wong
National Institute for Materials Science, Tsukuba, Japan

Christopher Y. H. Chao
The Hong Kong Polytechnic University, Hung Hom, Hong Kong, China

Satoshi Ishii
National Institute for Materials Science, Tsukuba, Japan

8.1 INTRODUCTION

Passive radiative cooling is a green technology without a carbon footprint that utilizes a sky-facing surface emitting thermal radiation through the bandwidth coincident with the infrared transparent atmospheric window lying within 8–13 μm of electromagnetic spectrum and self-preserving the temperature below ambient. It is deeply emerged in our daily life in which fog formation on land and plants under a nocturnal clear sky is a typical example. In 1978, with a titanium dioxide painted surface, it was the first time sub-ambient radiative cooling at daytime being reported in a scholarly research publication [1]. In 2014, thanks to escalating global concern on energy conservation and carbon neutrality, with a photonic radiative cooler composed of a bottom silver mirror and a top cascaded alternating silicon dioxide and hafnium dioxide 8–13 μm

DOI: 10.1201/9781003409090-8

thermal emitter, Raman et al.'s demonstration on daytime radiative cooling, realizing a temperature reduction of 5°C and a radiative cooling power of 40 W/m², drew intensive attention of the scientific community [2]. Then tremendous spectrally selective radiative coolers with improved thermal performance, broad materials selection, and scalable manufacturing feasibility were suggested within a few years [3–12]. They are expected to evolve numbers of engineering applications, especially for thermal and electrical systems. Aiming to review our latest developments on passive radiative cooling applications for thermal and electrical energy harvesting, this chapter is organized as follows. In Section 8.2 is a discussion of the analytical framework of thermo-photonic energy conversion for chilled water collection. Section 8.3 discusses one of the photo-thermoelectric energy conversion for electricity generation. In Section 8.4, is a discussion confronting the research challenge in passive radiative cooling. Last, Section 8.5 summarizes this chapter briefly. The advancement in radiative cooling materials is not discussed in this chapter because there have been a lot of published articles comprehensively reviewing the topic [13–15].

8.2 THERMO-PHOTONIC ENERGY CONVERSION FOR CHILLED WATER COLLECTION

A commercial chiller removes heat from chilled water via a vapor compression cycle, producing a cooling effect through a reversed Rankine cycle. A radiative-cooling-based chilled water system, simply integrable with a building's heating, ventilation, and air-conditioning system via a heat exchange interface, can substitute a part of the cooling load with a chiller and directly save electricity consumption. From the 1990s to 2000s, radiative-cooling-based chilled water systems were suggested and studied for nocturnal cooling capacity [16–21]. After Raman's succession in daytime radiative cooling, sub-ambient water cooling up to 5 °C during the daytime was demonstrated with a copper-tube-embedded aluminum plate exchanger [22]. Then a kilowatt scale radiative cooled cold collection system, also called RadiCool, was developed to chill water up to 10.6 °C at noon [23, 24]. Besides sensible cooling, water condensate can be harvested at the peak rate of 50 mL/m²-hr through latent cooling [25, 26]. Building energy simulations and modeling predicted energy saving in office buildings by 45–68 % relative to variable air volume HVAC system [27], cooling electricity saving in two-floor single-family houses by 26–46 % relative to split-type air conditioner [28], and indoor air temperature reduction up to 10 °C [29]. Unfortunately, chilling capacity of these radiative fluid cooling systems are bounded by a moderate radiative cooling power of 100 W/m². To efficiently utilize this new form of renewable energy resource, it is especially important to have a comprehensive understanding on their thermal and energy conversion performances, indicative by the fluid temperature reduction and energy conversion efficiency respectively.

Fluid temperature reduction denotes the fluid temperature difference before and after cooling, and energy conversion efficiency represents the ratio of enthalpy converted by the working fluid to the cooling effect harvestable from the sky [30]. To acknowledge the harvestable cooling effect, it is essential to consider energy balance of a radiative cooler subjected to a generic heat load q, which can be mathematically written as

$$P_{rad}\left(T_w\right) - P_{atm}\left(T_{amb}\right) - P_{sun} - h_c\left(T_{amb} - T_w\right) - \frac{q}{A} = 0, \quad (8.1)$$

where T_w is the surface temperature, T_{amb} is the ambient temperature, h_c is the coefficient of heat transfer between the surface and the environment, and A is the surface area. At an arbitrary temperature T,

$$P_{rad}\left(T\right) = \varepsilon\left(\lambda\right) I_{bb}\left(\lambda, T\right) d\lambda \cos\theta d\Omega, \quad (8.2)$$

is the radiative heat flux emitted by the radiative cooling surface,

$$P_{atm}\left(T\right) = \varepsilon\left(\lambda\right)\left[1 - \tau_{atm,0}\left(\lambda\right)^{1/\cos\theta}\right] I_{bb}\left(\lambda, T\right) d\lambda \cos\theta d\Omega, \quad (8.3)$$

is the radiative heat flux absorbed from the atmosphere, and,

$$P_{sun} = \varepsilon\left(\lambda\right) I_{AM1.5G}\left(\lambda\right)\cos\Psi d\lambda, \quad (8.4)$$

is the radiative heat flux absorbed from the sun, where $I_{bb}(\lambda, T) = 2h_p c^2/\lambda^5(e^{h_p c/\lambda k_b T} - 1)$ is the blackbody radiance, $c = 3 \times 10^8$ m/s is the speed of light, $h_p = 6.63 \times 10^{-34}$ J-s is the Planck's constant, $k_b = 1.38 \times 10^{-23}$ J/K is the Boltzmann constant, θ is the zenith angle of spherical coordinate system, Ω is the solid angle extending the upper hemisphere of spherical coordinate system, ψ is the zenith solar angle, λ is the wavelength,

$I_{AM1.5G}$ is the air mass 1.5 global solar radiance, ε is the spectral emissivity of radiative cooling surface, and $\tau_{atm,0}$ is the zenith atmospheric spectral transmittance.

When the surface temperature is raised to the ambient, the surface withstands the critical heat load that depends on ambient temperature only and equals the radiative cooling power in magnitude. Despite measuring the dischargeable electromagnetic energy, cooling power can be affected by materials properties, whereas harvestable cooling effect should be an intrinsic property of the sky as a heat dissipative thermal reservoir. Hence, cooling capacity should be determined by the ideal cooler capable in capturing the most cooling effect, performing as a spectrally selective blackbody emitter within 8–13 μm, but a perfect mirror elsewhere. Now energy conversion efficiency is well-defined as,

$$\eta_{th} = -\frac{\rho_f c_{p,f} Q_f \left(T_{f,c} - T_{f,h}\right)}{P_{net,ideal}\left(T_{amb}\right)}, \tag{8.5}$$

where ρ_f is the fluid density, $c_{p,f}$ is the specific heat capacity of fluid, Q_f is the flow rate of fluid, $T_{f,h}$ is the fluid temperature before cooling, $T_{f,c}$ is the fluid temperature after cooling, and $P_{net,ideal}(T_{amb})$ is the ideal radiative cooling power.

8.2.1 Energy Balance Model

To formulate the analytical framework for fluid temperature reduction and energy conversion efficiency, it is necessary to consider the 1-dimensional heat transfer model of a radiative fluid cooling system specified as follow. As shown in Figure 8.1(a) schematically, A rectangular channel of length l, width w, and height τ is engraved on a radiative cooler of area A. It connects two reservoirs at distinct temperatures of $T_{f,h}$ and $T_{f,c}$ respectively at the ends. Working fluid of density ρ_f and specific heat capacity $c_{p,f}$ drifts at a flow rate Q_f from the reservoir at $T_{f,h}$ to the one at $T_{f,c}$. Under uniform surface temperature T_w and adiabatic reservoir surfaces assumptions, an overall energy balance equation can be expressed as

$$P_{rad}\left(T_w\right) - P_{atm}\left(T_{amb}\right) - P_{sun} - h_c\left(T_{amb} - T_w\right) - \frac{\rho_f c_{p,f} Q_f}{A}\left(T_{f,h} - T_w\right)$$
$$\times \left(1 - e^{-\frac{h_{fw} w l}{\rho_f c_{p,f} Q_f}}\right) = 0, \tag{8.6}$$

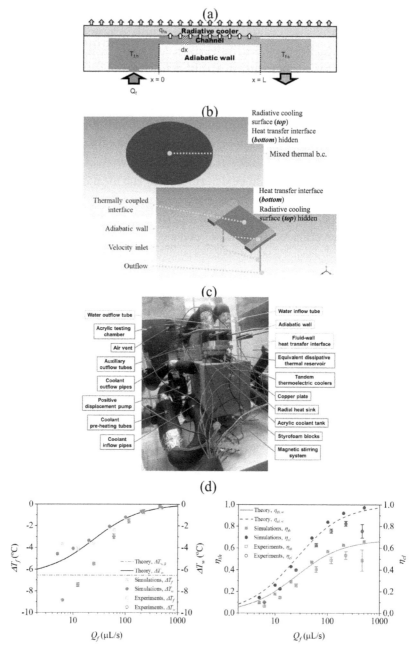

FIGURE 8.1 (a) One-dimensional heat transfer model; and (b) computational fluid dynamics simulation model for the passive radiative fluid cooling system; (c) a picture of the experimental setup for equivalent dissipative thermal reservoir experiment; a comparison of theoretical, simulated, and experimental results of (d) temperature reductions and efficiencies for flow rate between 2 μL/s and 1 mL/s;

(Continued)

(e)

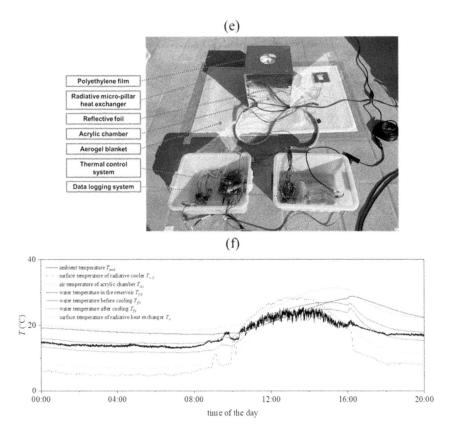

(f)

FIGURE 8.1 (Continued) (e) a picture of the experimental setup for outdoor field investigation on chilled water capacity; and (f) daily profiles of measured system temperatures during outdoor field investigation on chilled water capacity. Part (c) is reprinted from *International Journal of Heat and Mass Transfer*, **174** Wong RYM, Tso CY, Chao CYH, Corrected radiative cooling power measured by equivalent dissipative thermal reservoir method, 121341, Copyright (2021), with permission from Elsevier. Parts (a), (b) and (d) are reprinted from *Renewable Energy*, **180**, Wong RYM, Tso CY, Chao CYH, Thermo-radiative energy conversion efficiency of a passive radiative fluid cooling system, 700 – 711, Copyright (2021), with permission from Elsevier. (Parts (e) and (f) are reprinted from *International Journal of Heat and Mass Transfer*, **215**, Wong RYM, Tso CY, Fu SC, Chao CYH, Field demonstrated extended Graetzian viscous dissipative thermo-photonic energy conversion with a blended MgO/PVDF/PMMA coated glass-PDMS micro-pillar heat exchanger, 124520, Copyright (2023), with permission from Elsevier.)

where h_{fw} is the fluid-wall interfacial heat transfer coefficient. In this equation, fluid-wall interfacial heat transfer is entirely contributed by forced convection in the closed channel and, in this circumstance, h_{fw} is dependent on the frontal area and mode of heating only and solvable by classical heat transfer theory. Numerical values of Nusselt number $Nu_{fw, D}$, the dimensionless form of h_{fw} equaling $Nu_{fw, D} = h_{fw}D_H/\kappa_f$, where κ_f is the thermal conductivity of fluid and D_H is the hydraulic diameter, are available in many heat transfer textbooks like ref. [31], ranging between 2.5 and 8.2 for fully developed laminar channel flow. Radiative fluid cooling performance is optimal upon satisfying the criteria of $h_{fw}wl/\rho_f c_{p, f}Q_f >> 1$. In this circumstance, eq. (8.6) can be further simplified by considering the system subjected to a small perturbation in fluid flow from Q_f to $Q_f + dQ_f$. The direct consequence is two-folded. First, T_w is shifted by dT_w. Second, overall energy balance reacts in two aspects, in which the heat currents, P_{rad} and $h_c(T_{amb} - T_w)$, are altered in response to the change in T_w. Then the differential energy response equation can be worked out by subtracting the energy balance equations at two different system statuses. And, recognizing the boundary conditions that saturation temperature reduction ΔT_∞ equals the unloaded temperature reduction and saturation energy conversion efficiency, $\eta_{th, \infty}$ vanishes as $Q_f \to 0$, as well as $\Delta T_\infty = 0$ and $\eta_{th, \infty} = 1$ as $Q_f \to \infty$, analytical expressions for ΔT_∞ and $\eta_{th, \infty}$ can be obtained by integration as,

$$\Delta T_\infty = -\frac{P_{net}(T_{amb})}{\frac{\rho_f c_{p,f}Q_f}{A}\left(1+\frac{4E(T_{amb})\sigma T_{amb}^3 A}{\rho_f c_{p,f}Q_f}+\frac{h_c A}{\rho_f c_{p,f}Q_f}\right)}, \tag{8.7}$$

and,

$$\eta_{th,\infty} = \frac{P_{net}(T_{amb})}{P_{net,ideal}(T_{amb})\left(1+\frac{4E(T_{amb})\sigma T_{amb}^3 A}{\rho_f c_{p,f}Q_f}+\frac{h_c A}{\rho_f c_{p,f}Q_f}\right)}, \tag{8.8}$$

respectively, where $\sigma = 5.67 \times 10^{-8}$ W/m²-K⁴ is the Stefan Boltzmann constant and E is the ratio of the change in emitted radiative heat flux by a grey-body to a blackbody, and $P_{net}(T_{amb})$ is the radiative cooling power.

8.2.2 Computational Fluid Dynamics Simulations Model

Radiative fluid cooling performance can be demonstrated by computational fluid dynamics simulation. It involves two main procedures, including the meshing of computational domain and the iteration of discretized governing equations under specified boundary conditions sequentially. As shown in Figure 8.1(b), the simulation model involves not only the fluid advancing space, but also the substrate for radiative cooling materials deposition, interconnected by a heat transfer interface. The model is decomposed into a finite number of elements with local refinement at the boundary faces and interface. For fluid sub-domain, the governing equations are given by the mass, momentum, and energy conservation equations which take the form of

$$\nabla \cdot \mathbf{v} = 0, \tag{8.9}$$

$$\rho_f \left(\nabla \cdot \mathbf{v} \right) \mathbf{v} = -\nabla p + \nabla \cdot \tau, \tag{8.10}$$

and,

$$\rho_f c_{p,f} \left(\nabla \cdot \mathbf{v} \right) T = \kappa_f \nabla^2 T, \tag{8.11}$$

respectively for a single species, viscous and constant properties fluid, where $\tau = \mu_f (\nabla \mathbf{v} + \nabla \mathbf{v}^T)$ is the stress tensor, μ_f is the fluid viscosity, p is the hydrostatic pressure, $\mathbf{v}$ and $\mathbf{v}^T$ are the fluid velocity vector and transpose of fluid velocity vector. Fluid velocity and temperature are specified at the inlet. Outflow boundary condition, where the gradients of flow variables are vanished, is employed at the outlet. And no-slip hydrodynamic boundary condition and adiabatic thermal boundary condition are applied on all surrounding walls. For solid sub-domain, mass and momentum transports are negligible, and energy transport is driven by conduction only, where the heat conduction equation can be written as

$$\nabla \cdot \left(\kappa_s \nabla T \right) = 0, \tag{8.12}$$

where κ_s is the thermal conductivity of substrate. A mixed heat current, composed of a radiative current and a convective current, is specified at the radiative cooling surface. Thermally coupled wall condition is imposed at the fluid-wall heat transfer interface. An adiabatic wall boundary condition is set at the remaining walls. Finally, the governing equations are discretized by the second order upwind scheme, coupled with the SIMPLE algorithm, and solved by the finite element method. The relaxation factors are set in the range from 0.2 to 0.95. The discretized equations are iterated until the normalized residues are reduced to 10^{-5} or below for mass and momentum conservation equation, and 10^{-9} or below for energy conservation equation.

8.2.3 Equivalent Dissipative Thermal Reservoir Experiment

Furthermore, radiative fluid cooling performance can be illustrated by equivalent dissipative thermal reservoir experiment. It makes use of the strong linear dependence of net radiative heat exchange between the radiative cooling surface and the environment when surface temperature and ambient temperature are slightly different. As such, fluid cooling performance can be examined under the replicated radiative cooling effect established by a linear free buoyant stream between the dummy radiative fluid cooling system and the equivalent dissipative thermal reservoir. As shown in Figure 8.1(c), the equivalent dissipative thermal reservoir at the effective temperature, constructed by a surface with the same area of dummy radiative cooling surface, was conducted to the cold side of a thermo-electrical cooling system. A copper plate, capping a heat exchanger, was channeled to the hot side. Coolant, kept at a constant temperature, was chilled and circulated by the refrigerative chillers. It took away residual heat pumped by the thermoelectric coolers upon execution. The two systems were mounted on a distance and orientation adjustable platform to set up the designated thermal boundary conditions. More details on the equivalent dissipative thermal reservoir experiment can be referred to ref. [32].

Figure 8.1(d) compares the theoretical, simulated, and measured surface temperature reduction, fluid temperature reduction, and energy conversion efficiency. Saturation temperature reduction decreases with flow rate, whereas saturation energy conversion efficiency increases with flow rate. At small flow rate of 2 μL/s, temperature reduction approaches the unloaded value of 6.5 °C, but efficiency falls to 0 %. At a large flow rate of 1 mL/s, temperature reduction declines to 0 °C, but efficiency climbs to the ideal limit of 100 %. This inversed correlation can also be captured by simulation and experiment. Simulated surface and fluid temperature reductions come to the same point. And experimental results show that water can be chilled by 4.1 °C and cooling effect can be harvested by 212 mW, equivalent to the energy conversion efficiency of 14 %, whereas water can be weakly chilled by 1.5 °C and cooling effect can be harvested by 726 mW, equivalent to an elevated efficiency of 49 %. For flow rate over 20 μL/s, these tendencies are coherent with the theoretical prediction. For flow rate below 20 μL/s, simulated and experimental results are distinguishable from the analytical prediction. Despite the same simulated surface and fluid temperature reductions, the declining trends are not persistent with smaller flow rate and they plateau at 4.6 °C. Also, measured surface and fluid temperature reductions are not convergent. When the flow rate is 12.4 µL/s, fluid temperature reaches the bottom of 4.1 °C, higher than the saturation temperature reduction by 0.3 °C, but surface temperature reduction drops with flow rate continually, significantly lower than the saturation temperature reduction by 3 °C. As a consequence, energy conversion efficiency is lower than the saturation value.

8.2.4 Outdoor Field Investigation of Chilled Water Capacity

Furthermore, an outdoor field investigation of chilled water capacity was conducted with a wafer-sized radiative cooling blend coated glass-polydimethylsiloxane (glass-PDMS) micro-pillar heat exchanger [33]. The heat exchanger was composed of a micro-pillar patterned polydimethylsiloxane (PDMS) slab and a glass substrate. The PDMS slab was prepared by silicon stamping method [34, 35], in which the silicon master mask was fabricated by sequential micro-fabrication processes. PDMS, prepared by mixing the elastomer and curing agent in 10:1 was poured onto the patterned mask, baked in the oven for curing and lifted off from the mask after solidification. Lastly, a micro-porous radiative cooling blend was sprayed-coated on the opposite face. The blend was selected by recognizing the complementary thermal emissive property of poly(vinylidene-fluoride) and poly(methyl-methacrylate) through Maxwell-Garnett effective medium theory [36] and appraising the excellent solar reflective property of large energy bandgap dielectric materials [37–40]. Fourier transform mid-infrared spectrometry and UV/Vis/NIR spectrometry reveal high sky window emissivity and solar reflectivity.

As the experimental setup shown in Figure 8.1(e), the chilled water system, driven by a peristaltic pump, circulated water through the micro-pillar heat exchanger, water heating tube, and water reservoir at a constant flow rate of 6.3 µL/s. Besides, a radiative cooler with the same blend coated on a glass substrate was installed for radiative cooling power measurement. All components were well-insulated from conduction, convection, and radiation. Surface temperature of the radiative heat exchanger, water temperatures before and after cooling, as well as ambient temperature, were measured and their daily profiles were depicted in Figure 8.1(f). Beginning from sunset, the system arrived at different pseudo-steady temperatures at mid-night when the ambient temperature was 15.7 °C. Surface temperature of the radiative heat exchanger lay at 7.7°C, and it chilled water from 17.1°C to 14.5°C, equivalent to surface and water temperature reductions by 8.0°C and 2.6°C respectively. At noon of the second day, ambient temperature climbed to 23.0°C. Surface temperature of the radiative heat exchanger remained near ambient at 22.8°C, and it chilled water from 24.2°C to 22.8°C, equivalent to surface and water temperature reductions by 0.2°C and 1.4°C respectively. At nighttime, cooling power was measured by loading the blend-coated radiative cooler to ambient temperature. Just before the measurement, ambient temperature and surface temperature of the radiative cooler were 15.0°C and 6.0°C respectively. This denoted an unloaded surface temperature reduction of 9.0°C. During the measurement, ambient temperature declined slightly from 15.9°C to 14.1°C and surface temperature followed the same trend. Averaged temperatures differed by 0.3°C due to thermal control error. In this scenario, it gave a cooling power of 131 W/m² on per area basis. Meanwhile, the radiative heat exchanger arrived at a surface temperature of 9.4°C and chilled water by 2.3°C. Hence, estimated cooling efficiency with these figures was 5.9 %.

Compared to the temperature reduction and efficiency predicted by Equations (8.7) and (8.8), measured values are significantly lower even though the system satisfies the saturation criteria, perhaps because, at such a small flow rate, axial heat conduction, neglected in the one-dimensional heat transfer model, can be comparative to the convective and interfacial heat currents. The interplay among these heat currents causes a loss of interfacial heat transfer, but an increase in internal viscous dissipation. As a result, it degrades overall chilled water capacity by passive radiative cooling in this flow regime.

8.3 PHOTO-THERMOELECTRICAL ENERGY CONVERSION FOR ELECTRICITY GENERATION

Photo-thermoelectrical generation is an equally important engineering application of passive radiative cooling technology. At daytime, in-site photo-electrical energy conversion can be simply realized by a photovoltaic cell. Nowadays, with a new anti-reflection coating, a concentrated multi-junction tandem solar cell made of III-V compounded semiconductors recorded the state-of-art solar-electrical energy conversion efficiency of 47.6 % [41]. At nighttime, similar direct renewable energy conversion technology was overlooked until recent attempts on photo-thermoelectricity generation enabled by radiative cooling. It makes use of the heat dissipative radiative cooling coating to set up a temperature gradient between the hot and cold sides of a Peltier module and extract electrical work between the ambient and universe. Passive radiative-cooling based thermoelectrical energy harvesting has potential applications in various aspects, including building space cooling [42, 43], and powering wearable electronics [44, 45].

8.3.1 Energy Balance Model

Theoretically, system thermal performance, indicated by cold-sided temperature $T_{TE,c}$ and hot-sided temperature $T_{TE,h}$, can be acknowledged by solving the energy balance equations for the Peltier module of area A, comprising N p-n junctions with an individual Seebeck coefficient S_{pn}, thermal conductance K_{pn} and electrical resistance R_{pn}, connected to an external load of resistance R_e. For the cold side,

$$P_{rad}\left(T_{TE,c}\right) - P_{atm}\left(T_{amb}\right) - P_{sun} - h_c\left(T_{amb} - T_{TE,c}\right) - P_{cond}\left(T_{TE,h}, T_{TE,c}\right)$$
$$- P_{seebeck}\left(T_{TE,c}\right) - P_{joule} = 0 \tag{8.13}$$

and, for the hot side,

$$-h_c\left(T_{amb} - T_{TE,h}\right) + P_{cond}\left(T_{TE,h}, T_{TE,c}\right) + P_{seebeck}\left(T_{TE,h}\right) - P_{joule} = 0, \tag{8.14}$$

where $P_{cond} = NK_{pn}(T_{TE,h} - T_{TE,c})/A$ is the conductive heat flux, $P_{seebeck} = NS_{pn}TI/A$ is the Seebeck effect, $P_{joule} = NI^2R_{pn}/2A$ is the joule heating, and $I = NS_{pn}(T_{TE,h} - T_{TE,c})/(NR_{pn} + R_e)$ is the thermoelectrical current. In eq. (2.1), P_{rad}, P_{atm} and P_{sun} are given by Equations (8.2), (8.3) and (8.4) respectively and $h_c(T_{amb} - T_{TE,c})$ is the non-radiative heat load from the environment. Hence, system electrical performance, denoted by thermoelectrical power output P_e, is coupled with the difference in $T_{TE,h}$ and $T_{TE,c}$ via I, and given by [46, 47]

$$P_e = \frac{N^2 S_{pn}^2 \left(T_{TE,h} - T_{TE,c}\right)^2 R_e}{\left(NR_{pn} + R_e\right)^2}. \tag{8.15}$$

An outdoor field investigation of the black-paint coated commercial thermoelectric module realized a two-sided temperature difference up to 2 °C and generated a thermoelectrical power of 25 mW/m² at the maximum power point, in which the field investigative result validates the theoretical model [48]. Predicted thermoelectrical performance is divergent and highly sensitive to the input system parameters, in which Zhao et al. estimated a moderate output power density of 291 mW/m² [47], whereas Fan et al. forecasted a much larger optimal value of 2.2 W/m² [46].

8.3.2 Laboratory Testing on Thermo-Electricity Generation

A laboratory testing comparing thermo-electricity generation by Peltier modules combined with a selective thermal emitter and a blackbody thermal emitter was set up for duplicating thermo-electricity generation [49]. As shown in Figure 8.2(a), the selective thermal emitter was an e-beam evaporated 100 nm aluminum film on a glass with the glass side facing top, and the blackbody thermal emitter was a blackbody paint-sprayed glass with the blackbody paint facing top. Figure 8.2(b) shows their spectral emissivity and reflectivity from visible to mid-infrared wavelengths.

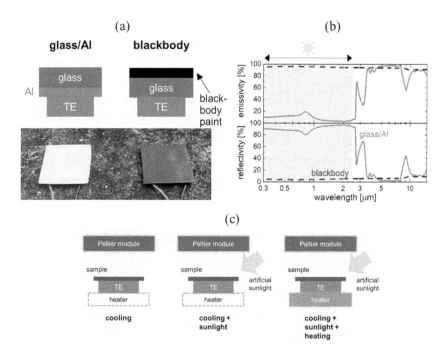

FIGURE 8.2   (a) Pictures of the selective and blackbody thermal emitters; (b) measured spectral emissivity and reflectivity of the selective and blackbody thermal emitters from visible light to mid-infrared wavelengths; (c) An illustrative diagram on the experimental conditions for laboratory testing;

(Continued)

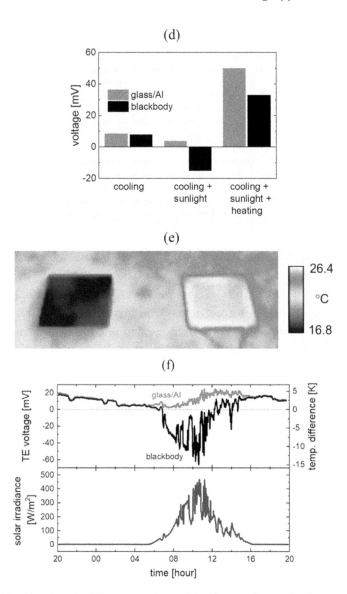

FIGURE 8.2 (Continued) (d) a comparison of the thermoelectrical voltage generated under abovementioned three experimental conditions; (e) thermograph; and (f) measured temperature difference and thermoelectrical voltage of the selective and blackbody thermal emitters under field investigation. (Reprinted from *Applied Physics Letters*, **117**, Ishii S, Dao TD, Nagao T, Radiative cooling for continuous thermoelectric power generation in day and night, 013901, with permission from AIP Publishing.)

The glass/aluminum emitter has a high reflectivity in the optical range, but a high emissivity in the mid-infrared range.

A blackbody-painted Peltier module and a solar simulator were used to imitate the universe and the sun. As illustrated in Figure 8.2(c), three experimental conditions, simulating a standalone radiative cooling effect, combined radiative cooling and solar heating effect, and combined radiative cooling, solar heating, and conductive heating effect, were trialed. Figure 8.2(d) compares the thermoelectrical voltage generated under abovementioned three experimental conditions. Where there is radiative cooling only, the two samples generated similar voltage. When there are both radiative cooling and solar heating, the one with selective thermal emitter generated positive voltage, whereas the one with the blackbody thermal emitter generated negative voltage with a larger absolute magnitude. Solar heating overwhelmed radiative cooling, resulting in a larger temperature difference and voltage. However, when the samples were heated at the bottom, the former generated a larger voltage than the latter. Top-sided radiative cooling and bottom-sided conductive heating maximized the temperature difference and voltage. When the bottom-sided temperature increased from room temperature to 50 °C, thermoelectrical voltage of the one with a selective thermal emitter increased from 3.8 to 50 mV, whereas the one with the blackbody thermal emitter changed from −15 to 33 mV only. This indicated that a selective thermal emitter could take advantage of bottom-sided heating from waste heat in practice for enhanced radiative-cooling based thermo-electricity generation.

8.3.3 Outdoor Field Investigation of Thermo-Electricity Generation

Then, an outdoor field investigation was conducted to investigate the performance. The selective thermal emitter was radiatively cooled all the time, and the top of the thermoelectricity generator is always cooler than the bottom, thus maintaining continuous thermoelectricity generation. In contrast, the black-paint emitter has a high emissivity across the entire spectrum. The blackbody thermal emitter was radiatively cooled at nighttime but heated by sunlight at daytime. Thus, after sunrise, the top surface temperature increased gradually, and reached the same bottom face temperature gradually. Without vanished voltage, continuous thermoelectricity generation acted as the overwhelming advantage of radiative-cooling based electrical energy harvesting. Figure 8.2(e) shows the thermograph of two devices under field investigation taken in day. As can be seen, the radiative cooling surface was at a temperature lower than the background, whereas the blackbody was at a higher temperature. Figure 8.2 (f) shows the measured temperature difference and thermoelectrical voltage during field investigation. Note that the temperature difference is positive/negative when the emitter surface is cooler/hotter than the ambient temperature At nighttime, top-to-bottom temperature differences were approximately 2–4 °C for both devices. Consequently, they produced similar thermoelectrical voltage up to 20 mV. At daytime, the temperature difference arrived at 5 °C for the selective thermal emitter, whereas, with a higher top face temperature due to solar heating, it reached a larger value of −15 °C for the blackbody thermal emitter. Hence, the thermoelectricity generator installed by the blackbody thermal emitter generated a larger voltage, up to 60 mV, because solar heating by the blackbody absorber is stronger than radiative cooling by the selective radiative cooler. Moreover, from 1pm to 5pm, the sky was cloudy and inferred from the lower solar irradiance. During this period, the device with a selective thermal emitter recorded a larger thermoelectrical voltage than the one with a blackbody thermal emitter. Furthermore, it generated nearly constant thermoelectrical voltage regardless of weather change.

Later, it was shown that reducing parasitic losses, controlling emitter area and thermal resistance of the thermoelectric generator, and stacking multiple thermoelectric generators are all effective ways to boost the power density. Also, a measured power density exceeding 100 mW/m², representing over 2-fold improvement over the previous results, was demonstrated experimentally [50]. Also, a new kind of thermoelectrical energy converter based on the spin Seebeck effect, in which the temperature gradient and the thermoelectrically generated electric field are perpendicular, was suggested for energy harvesting from solar heating and radiative cooling simultaneously [51]. The Spin Seebeck effect induced voltage is proportional to the length of the device, which is perpendicular to the temperature gradient. This means that voltage and power can be increased by simply elongating the device length without forming multitude of serial p-n junctions, as is the case with a conventional thermoelectric device. And, this simplifies the device architecture. A prototype, comprising paramagnetic gadolinium gallium garnet substrate, ferrimagnetic yttrium iron garnet insulator, paramagnetic platinum metal, and blackbody paint light absorber, demonstrated the simultaneous harvesting of radiative cooling and solar heating in the outdoors.

8.4 RESEARCH CHALLENGE IN PASSIVE RADIATIVE COOLING

Geographical variation in the passive radiative cooling performance poses one of the top research challenges in low latitude hot and humid regions. The radiative cooling resource map for the contiguous United Stated showed that the southwestern area had the highest cooling potential of 70 W/m², whereas the southeastern had the lowest potential of 30 W/m² [52]. Similar maps for China identified that the northwestern area had the highest cooling potential of 70–90 W/m², whereas the southeastern had the lowest potential of 10–40 W/m² [53, 54]. An investigation of the impact of humidity, cloudiness, and aerosol concentration on radiative cooling performance compared the cooling potential at Stanford and Hong Kong, where the estimated values were 61 W/m² and 25 W/m² respectively due to climatic difference [55]. Besides, higher solar intensity in Singapore, where the predicted cooling power limit was 30 W/m², was identified as the major cause of degraded radiative cooling performance [56]. As such, despite plenty of groundbreaking field investigative reports from North America, some comparative studies conducted elsewhere failed to achieve sub-ambient daytime radiative cooling [57–61]. In Shanghai, where the ambient temperature and relative humidity were above 24 °C and 50 % respectively, nano-particle based solar reflecting thermal radiators were tested, but they remained 3–10 °C above ambient at daytime [57]. In the subtropical city, Hong Kong, Raman's photonic radiative cooler was tested in vacuum and non-vacuum enclosures, but none can replicate sub-ambient daytime radiative cooling [2, 58]. Then a modified titanium oxide photonic radiative cooler and a bio-inspired polymeric radiative cooler were tested with and without shade, and only the shaded ones accomplished sub-ambient daytime radiative cooling [59, 60]. In the tropical country, Singapore, Raman's photonic radiative cooler and an enhanced specular reflective film were assessed. It was concluded that high solar intensity and humidity counteracted the radiative cooling effect [2, 61].

8.4.1 Empirical Sky Temperature Models

Contrary to mainstream deterministic energy balance-based analysis of the variance in radiative cooling performance arising from climatic factors, a new approach by means of probabilistic regression modeling was suggested to establish the correlation between radiative cooling performance and corresponding weather conditions [62]. It is advantageous in tolerancing the uncertainties arising from time varying and uncontrolled atmospheric uncertainties abundant in field investigation. Meteorological variables like ambient temperature, relative humidity, and cloudiness, quantifying the downwelling atmospheric thermal radiation from the sky, can be lumped into a single parameter of sky temperature. Sky temperature T_{sky} can be viewed as the equivalence of atmospheric thermal intensity I_{sky} in absolute temperature scale, convertible by the Stefan-Boltzmann equation of $I_{sky} = \sigma T_{sky}^4$, where $\sigma = 5.67 \times 10^{-8}$ W/m²-K⁴ is the Stefan-Boltzmann constant. And sky emissivity ε_{sky} is defined by T_{sky} via $\varepsilon_{sky} = (T_{sky}/T_{amb})^4$. As ε_{sky} ranges from 0 to 1, T_{sky} is always lower than T_{amb}. ε_{sky} can be measured by a pyrometer experimentally, simulated by an atmospheric radiative transfer model numerically, and a sky emissivity model empirically. A pyrometer produces an output voltage by scanning in situ infrared radiance within 4.5–100 μm, but it requires careful calibration to eliminate background radiation from buildings and vegetations. An atmospheric radiative transfer model evaluates the atmospheric spectral emissivity within 0.2–100 μm line-by-line upon comprehensive specification of absorbing gases, aerosol, water vapor, cloud characteristics, vertical temperature, and humidity profiles, as well as various secondary atmospheric variables [63, 64]. In contrast, an empirical model is rather simple, correlating ε_{sky} with fundamental meteorological variables measurable by the observatory.

Since the 1910s, plenty of sky temperature models, falling into two categories regarding clear and cloudy skies, have been suggested. Under a clear sky, suspended water vapors act as the primary source of downwelling atmospheric thermal radiation. In 1932, Brunt formulated one of the earliest clear sky temperature models, expressing clear sky emissivity $\varepsilon_{sky,c}$ as

$$\varepsilon_{sky,c} = a_1 + a_2 p_w^{\frac{1}{2}}, \tag{8.16}$$

where $a_1 = 0.52$ and $a_2 = 0.065$ are the empirical constants, and p_w is the vapor pressure [65]. Afterwards, various parametric forms were suggested, and published models were also recalibrated. These reports revealed the difficulty in universal sky temperature modeling because the models were established with localized and biased meteorological data drawn from one or several weather stations. A recent revisit on this topic might have analyzed the most comprehensive meteorological and radiation data, collected from seven stations of the Surface Radiation Budget Network in the United States, located at Goodwin Creek (Mississippi), Bondville (Illinois), Penn State University (Pennsylvania), Fort Peck (Montana), Sioux Falls (South Dakota), Boulder (Colorado), and Desert Rock (Nevada). The meteorological dataset, covering climatic diversity in the northern hemisphere, were used to recalibrate Brunt's model with renewed empirical constants of $a_1 = 0.62$ and $a_2 = 0.056$ [66]. Under a cloudy sky, clusters of liquid phase water and solid phase ices absorb and emit longwave radiation more vigorously than gaseous phase vapors. To cater for additional heat load imposed by clouds, clear sky emissivity ought to be corrected by a factor regarding cloud fraction. A new empirical form of cloudy sky emissivity ε_{sky} was suggested as

$$\varepsilon_{sky} = \varepsilon_{sky,c} \left(1 - b_1 f_c^{b_2}\right) + b_3 f_c^{b_4} \phi^{b_5}, \tag{8.17}$$

where $b_1 = 0.78$, $b_2 = 1$, $b_3 = 0.38$, $b_4 = 0.95$ and $b_5 = 0.17$ are the empirical constants, ϕ is the relative humidity, and f_c is the cloud fraction [66]. In the expression, $\varepsilon_{sky,c}$ can be calculated with recalibrated Brunt's clear sky emissivity by Equation (3.1). When $f_c = 0$, it represents the clear sky condition, and ε_{sky}, given by Equation (3.2), is reduced to $\varepsilon_{sky,c}$. Hence, cloudy sky temperature model is applicable for all sky conditions.

8.4.2 Correlations Between Sky Temperature Difference and Surface Temperature Reduction

Outdoor field investigations were conducted for photonic radiative and polymeric radiative coolers. For each kind of radiative cooler, one was shadowed by an external shade, and one was exposed to direct sunlight. During the investigations, surface temperatures and site ambient temperature were measured. Meanwhile, ambient temperature, relative humidity, cloud fraction, and solar intensity were collected from neighboring weather stations.

In total, 70 sets of field investigative results were gathered at nighttime, whereas 35 sets were gathered at the peak solar radiance. Then temperature measurements were time averaged. And Figure 8.3(a)–(d) shows the scatterplots of the response variable of surface temperature reduction against predictor variable of sky temperature difference for different materials at nighttime. Surface temperature reduction, as a cardinal indicator of radiative cooling performance, is a reasonable selection for the response variable. Also, sky temperature difference, which is the difference between

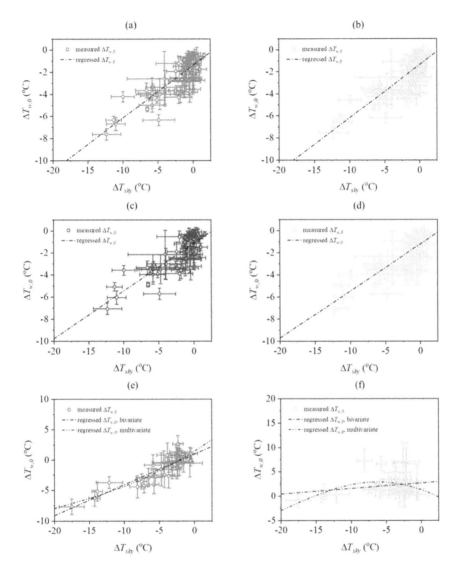

FIGURE 8.3 Scatterplots of surface temperature reduction at nighttime against sky temperature difference and bivariate regression lines for (a) shaded polymeric radiative cooler; (b) unshaded polymeric radiative cooler; (c) shaded photonic radiative cooler; and (d) unshaded photonic radiative cooler. Scatterplots of surface temperature reduction at daytime against sky temperature difference, bivariate and multi-variate regression lines for (e) shaded polymeric radiative cooler; (f) unshaded polymeric radiative cooler;

(Continued)

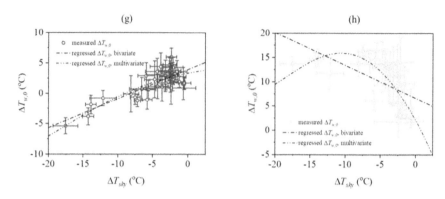

FIGURE 8.3 (Continued)   (g) shaded photonic radiative cooler; and (h) unshaded photonic radiative cooler. (Reprinted from *Renewable Energy*, **211**, Wong RYM, Tso CY, Jeong SY, Fu SC, Chao CYH, Critical sky temperatures for passive radiative cooling, 214–226, Copyright (2023), with permission from Elsevier.)

sky temperature and ambient temperature, can be an appropriate choice for the predictor variable. It is because, from an energy balance consideration, net radiative heat exchange by the radiative cooler results from outfluxing emission proportional to the fourth power of surface temperature and inflowing absorption proportional to the fourth power of sky temperature. For small differences among ambient temperature, sky temperature, and surface temperature, they can be expressed as a Taylor expansion about ambient temperature, and thus, net radiative heat exchange can be scaled with a single, lumped meteorological variable of sky temperature difference. From a statistical perspective, each data-pair scatters about the best fitted line. Supposed to be linear, the best fitted lines, based on optimal intercept coefficient and slope coefficient, are plotted in the same figure for reference. Variance of surface temperature reduction and R^2 value estimate the reliability of the regression model, where the former represents the amount of variability inherent in the regression model, and the latter interprets the proportion of variation in surface temperature reduction predictable from the sky temperature difference. The variances after normalization are approximately the same of -0.18 for all specimens, which means climatic factors impact on expected radiative cooling performance, but do not alter the random deviation, and R^2 values lie within 0.63 and 0.7, which denotes a substantial linear correlation between surface temperature reduction and sky temperature difference for nighttime radiative cooling performance.

At daytime, solar absorption imposes an additional heat load on the radiative coolers. Figure 8.3(e)–(h) plots the corresponding scatterplots under the peak solar intensity. Repeating the bivariate linear regression model for data analysis, variances and R^2 values lie within statistically valid ranges for shaded radiative coolers, whereas they appear as tremendous departures between measured datasets and regressed equations, and associated random errors are magnified enormously, but R^2 values are reduced dramatically, questioning the linear coupling between sky temperature difference and surface temperature reduction. Therefore, it is essential to revise the model by introducing an extra predictor variable regarding solar heat load. Obviously, sky temperature difference and solar intensity are not mutually independent variables, but solar intensity should be regarded as a single variable function of sky temperature difference, and its explicit form ought to be pre-determined otherwise. Beer-Lambert law states that [67, 68] spectral radiative extinction traversing a thin layer of medium is proportional to local spectral intensity, number density and extinction cross section of extinctive particles, as well as medium thickness. For the atmosphere composed of multi-component gases and suspended particulates, the contribution of extinction cross section by each species is additive and the integral form of Beer-Lambert law can be written as

$$I(\lambda,s) = I(\lambda,0)e^{-\int \sum_s \sum_j K_j(\lambda)N_j(s')ds'}. \tag{8.18}$$

Constituent gases, like nitrogen, oxygen, and argon, occupy permanent fractions in the atmosphere, whereas suspended particulates, like aerosols, water vapors, and clouds, are time and space varying in concentration. Complicated extinction mechanism and comprehensive specification of number density and extinction coefficient for each atmospheric constituent throughout the optical path do not facilitate practical implementation of Beer-Lambert law. In many circumstances, it is essential to recognize these changes with respect to climatic parameters like sky temperature rather than their absolute values. Because of a small variation in sky temperature, the term, $\sum_j K_j N_j$, in the Beer-Lambert equation at any sky temperature can be expressed as a Taylor's expansion about a reference sky temperature. And the contributions from water vapors and clouds become the only terms dynamic with sky temperature. Further neglecting spatial variations in extinction cross-sections of vapors and clouds yields the territorial solar intensity written as $I_{sun}\left(\Delta T_{sky}\right)=I_{sun}^{(0)}e^{-\left(\Delta T_{sky}-\Delta T_{sky}^{(0)}\right)\int_{S}\left(K_v\partial N_v/\partial T_{sky}+K_c\partial N_c/\partial T_{sky}\right)ds'}$, where $I_{sun}^{(0)}$ represents the solar intensity at $T_{sky}^{(0)}$, which can be as high as 1140 W/m^2 at low-to-mid altitude areas [69]. Admittedly, a constant extinction cross-section assumption may not stand because water vapors are spectrally selective solar absorbers, in which the atomic arrangement of water molecule permits three fundamental vibration modes of symmetry, bending, and anti-symmetry, responsible for multiple solar and near-infrared absorption bands at 940 nm, 1.1 μm, 1.38 μm and 1.87 μm [70]. The integral, $-\int_S(K_v\partial N_v/\partial T_{sky}+K_c\partial N_c/\partial T_{sky})ds'$, denotes the exponential declining rate of solar intensity with respect to sky temperature difference. In a semi-empirical treatment, it can be determined by surveying the historical meteorological data pairs of the peak solar intensity against the sky temperature difference, and appraising the slope at any arbitrary sky temperature difference is supposed to be the linear interpolation of two limiting characteristic rates. Hence, the integral can be simplified as $\gamma_1+2\gamma_2\Delta T_{sky}$, where γ_1 and γ_2 are the empirical model constants. They can be determined by minimizing the root-of-squared error between collected dataset and modeling equation. as such, $I_{sun}\left(\Delta T_{sky}\right)=I_{sun}^{(0)}e^{\gamma_0+\gamma_1\Delta T_{sky}+\gamma_2\Delta T_{sky}^2}$ and the multivariate regression equation becomes $\Delta T_{w,0}=\hat{\beta}_0+\hat{\beta}_1\Delta T_{sky}+\hat{\beta}_2 e^{\gamma_0+\gamma_1\Delta T_{sky}+\gamma_2\Delta T_{sky}^2}$. Compared to bivariate regression model, it provides a better fit with scattered data pairs in Figure 8.3(f)–(j), features decreased variances, as well as increased R^2 values, and improves overall interpretability of the statistical model. For shaded radiative coolers, variances range from 1.1 °C to 1.4 °C and R^2 values range from 0.69 to 0.79. The refinement is the least notable because, without the action of direct solar illumination, the bivariate regression model has rationalized radiative cooling performance. For exposed radiative coolers and silicon wafer, the advancement is more significant, revealing the crucial role of solar heat load on cooling performance. Even the polymeric radiative cooler feature reasonably high solar reflectivity variance is reduced to 2.0 °C, and R^2 value is more than doubled, valuing 0.11. The small random uncertainty affirms the reliability of the regression model, whereas the feeble correlation stems from the counter-interaction of radiative cooling and solar heating. For photonic radiative cooler, variance is 3.0 °C and R^2 value is 0.58. The multivariate regression model reduces the random uncertainty but reinforces the correlation between sky temperature difference and surface temperature reduction.

Under a subtropical hot and humid climate, sub-ambient passive radiative cooling is possible providing reconcilable materials with sky window emissivity and solar reflectivity higher than the benchmarked polymeric radiative cooler. A few alternative solution strategies to subtropical and tropical radiative cooling were also proposed. Providing external shading can be the simplest way to block incoming solar radiation and lower surface temperature [59, 60]. In Hong Kong, a radiative cooler with superior spectral selectivity, comprising a solution-derived silicon oxynitride layer sandwiched between a reflective substrate and a self-assembly monolayer of silicon dioxide microspheres narrowband emitter, realized sub-ambient cooling of up to 5 °C in autumn and 2.5 °C in summer [71]. In Singapore, a switchable solar heater radiative cooler with engineered porous structure, enabling the device to serve as an efficient solar reflector and infrared emitter in dry state, as well as an efficient solar heater in wet state, yielded a nighttime cooling power of 61.2 W/m² and daytime heating power of 720 W/m² [72]. These strategies create a new opportunity for the development of novel and multi-functional radiative cooling materials.

8.5 SUMMARY AND CONCLUSIONS

This chapter reviews passive radiative cooling applications in thermal and electrical energy harvesting. They are important because of plenty of smart and green technological applications toward a carbon neutral built environment. For thermal energy harvesting, chilled water collection by circulating water through a radiative heat exchanger is the simplest form, but it raises critical concerns on thermal and energy conversion performances. A study on a passive radiative fluid cooling system in a controlled environmental facility revealed that water temperature reduction and energy conversion efficiency are always inversely correlated, in which temperature reduction increases with decreasing flow rate but efficiency increases with increasing flow rate. This poses a fundamental difficulty in collecting chilled water in an energy efficient manner. For electrical energy harvesting, thermoelectricity by creating temperature difference through heat dissipation with radiative cooling materials at the cold side can be generated continuously day and night. It demands on the development of high-performance radiative cooling-based thermoelectricity generator delivering power density approaching the theoretical upper limit. However, it must be emphasized that performance of radiative cooling-based devices and systems rely on the weather conditions heavily. Sky temperature modeling unveils that, the higher the thermal emissivity, the larger is the surface temperature reduction at nighttime, and heavy solar heat load can be absorbed by the radiative coolers at daytime even they feature reasonably high solar reflectivity. Therefore, hot and humid climates are not favorites for passive radiative cooling. This complicates the applications in tropical and subtropical regions, where the cooling demand is the heaviest, though a few overcoming schemes were suggested. Nonetheless, advancement in passive radiative cooling materials and applications will not be stopped and will continuously steer a new shape of our community and society.

REFERENCES

[1] Harrison, A. W., Walton, M. R. (1978). Radiative cooling of TiO_2 white paint. *Solar Energy*, **20**, 185–188.

[2] Raman, A. P., Anoma, M. A., Zhu, L., Rephaeli, E., Fan, S. (2014). Passive radiative cooling below ambient air temperature under direct sunlight. *Nature*, **515**, 540–544.

[3] Chen, Z., Zhu, L., Raman, A., Fan, S. (2016). Radiative cooling to deep sub-freezing temperatures through a 24-h day-night cycle. *Nature Communications*, **7**, 13729.

[4] Kou, J., Jurado, Z. Chen, Z., Fan, S., Minnich, A. J. (2017). Daytime radiative cooling using near-black infrared emitters. *ACS Photonics*, **4**, 626–630.

[5] Hossain, M. M., Jia, B., Gu, M. (2015). A metamaterial emitter for highly efficient radiative cooling. *Advanced Optical Materials*, **3**, 1047–1051.

[6] Zou, C., Ren, G., Hossain, M. M., Nirantar, S., Withayachumnankul, W., Ahmed, T., Bhaskaran, M., Sriram, S., Gu, M., Fumeaux, C. (2017). Metal-loaded dielectric resonator metasurfaces for radiative cooling. *Advanced Optical Materials*, **5**, 1700460.

[7] Yu, N., Mandal, J., Overvig, A., Shi, N. N. (2016). Systems and Methods for Radiative Cooling and Heating, US patent: WO2016205717A1.

[8] Zhai, Y., Ma, Y., David, S. N., Zhao, D., Lou, R., Tan, G., Yang, R., Yin, X. (2017). Scalable manufactured randomized glass-polymer hybrid metamaterial for daytime radiative cooling. *Science*, **355**, 1062–1066.

[9] Mandal, J., Fu, Y., Overvig, A. C., Jia, M., Sun, K., Shi, N. N., Zhou, H., Xiao, X., Yu, N., Yang, Y. (2018). Hierarchically porous polymer coatings for highly efficient passive daytime radiative cooling. *Science*, **362**, 315–319.

[10] Peoples, J., Li, X. Y., Lv, Y. B., Qiu, J., Huang, Z. F., Ruan, X. L. (2019). A strategy of hierarchical particle sizes in nanoparticle composite for enhancing solar reflection. *International Journal of Heat Mass Transfer*, **131**, 487–494.

[11] Li, X., Peoples, J., Huang, Z., Zhao, Z., Qiu, J., Ruan, X. (2020). Full daytime sub-ambient radiative cooling in commercial-like paints with high figure of merit. *Cell Reports Physical Science*, **10**, 100221.

[12] Li, X., Peoples, J., Yao, P., Ruan, X. (2021). Ultrawhite $BaSO_4$ paints and films for remarkable daytime subambient radiative cooling, *ACS Applied Mater. Interfaces*, **13**, 21733–21739.

[13] Zhao, B., Hu, M., Ao, X., Chen, N., Pei, G. (2019). Radiative cooling: A review of fundamentals, materials, applications, and prospects. *Applied Energy*, 236, 489–513.

[14] Sun, J., Wang, J., Guo, T., Bao, H., Bai, S. (2022). Daytime passive radiative cooling materials based on disordered media: A review. *Solar Energy Materials and Solar Cells*, **236**, 111492.

[15] Xie, B., Liu, Y., Xi, W., Hu, R. (2023). Colored radiative cooling: Progress and prospects. *Materials Today Energy*, **34**, 101302.

[16] Ali, A. H. A., Taha, I. M. S., Ismail, I. M. (1995). Cooling of water flowing through a night sky radiator. *Solar Energy*, **55**, 235–253.

[17] Erell, E., Etzion, Y. (1999). Analysis and experimental verification of an improved cooling radiator. *Renewable Energy*, **16**, 700–703.

[18] Saitoh, T. S., Fujino, T. (2001). Advanced energy-efficient house (Harbeman house) with solar thermal photovoltaic and sky radiation energies (Experimental results). *Solar Energy*, **70**, 63–77.

[19] Meir, M. G., Rekstad, J. B., Lovvik, O. M. (2002). A study of a polymer-based radiative cooling system. *Solar Energy*, **73**, 403–417.

[20] Dimoudi, A., Androutsopoulos, A. (2006). The cooling performance of a radiator based roof component. *Solar Energy*, **80**, 1039–1047.

[21] Tevar, J. A. F., Castano, S., Marijuan, A. G., Heras, M. R., Pistono, J. (2015). Modelling and experimental analysis of three radioconvective panels for night cooling. *Energy and Buildings*, **107**, 37–48.

[22] Goldstein, E. A., Raman, A. P., Fan, S. (2017). Sub-ambient non-evaporative fluid cooling with the sky. *Nature Energy*, **2**, 17143.

[23] Zhao, D., Aili, A., Zhai, Y., Lu, J., Kidd, D., Tan, G., Yin, X., Yang, R. (2019). Subambient cooling of water: Toward real-world applications of daytime radiative cooling. *Joule*, **3**, 111–123.

[24] Aili, A., Zhao, D., Lu, J., Zhai, Y., Yin, X., Tan, G., Yang, R. (2019). A kw-scale, 24-hour continuously operational, radiative sky cooling system: Experimental demonstration and predictive modelling. *Energy Conversion and Management*, **186**, 586–596.

[25] Zhou, M., Song, H., Xu, X., Shahsafi, A., Qu, Y., Xia, Z., Ma, Z., Kats, M. A., Zhu, J., Ooi, B. S., Gan, Q., Yu, Z. (2021). Vapor condensation with daytime radiative cooling. *Proceedings of National Academy of Science*, **118**, 2019292118.

[26] Haechler, I., Park, H., Schnoering, G., Gulich, T., Rohner, M., Tripathy, A., Milionis, A., Schutzius, T. M., Poulikakos, D. (2021). Exploiting radiative cooling for uninterrupted 24-hour water harvesting from the atmosphere. *Science Advances*, **7**, 3978.

[27] Wang, W., Fernandez, N., Katipamula, S., Alvine, K. (2018). Performance assessment of a photonic radiative cooling system for office buildings. *Renewable Energy*, **118**, 265–277.

[28] Zhang, K., Zhao, D., Yin, X., Yang, R., Tan, G. (2018). Energy saving and economic analysis of a new hybrid radiative cooling system for single-family houses in the USA. *Applied Energy*, **224**, 371–381.

[29] Jeong, S. Y., Tso, C. Y., Zouagui, M., Wong, Y. M., Chao, C. Y. H. (2018). A numerical study of daytime passive radiative coolers for space cooling in buildings. *Building Simulation*, **11**, 1011–1028.

[30] Wong, R. Y. M., Tso, C. Y., Chao, C. Y. H. (2021). Thermo-radiative energy conversion efficiency of a passive radiative fluid cooling system. *Renewable Energy*, **180**, 700–711.

[31] Incropera, F. P., Dewitt, D. P., Bergman, T. L., Lavine, A. S. (2006). *Fundamentals of Heat and Mass Transfer* (6th edition), John Wiley & Sons, United States of America.

[32] Wong, R. Y. M., Tso, C. Y., Chao, C. Y. H. (2021). Corrected radiative cooling power measured by equivalent dissipative thermal reservoir method. *International Journal of Heat and Mass Transfer*, **174**, 121341.

[33] Wong, R. Y. M., Tso, C. Y., Fu, S. C., Chao, C. Y. H. (2023). Field demonstrated extended Graetzian viscous dissipative thermo-photonic energy conversion with a blended MgO/PVDF/PMMA coated glass-PDMS micro-pillar heat exchanger, *International Journal of Heat and Mass Transfer*, **215**, 124520.

[34] Xia, Y., Whitesides, G. M. (1998). Soft lithography. *Annual Review of Materials Science*, **28**, 153–184.

[35] Whitesides, G. M., Ostuni, E., Takayama, D., Jiang, X., Lngber, D. E. (2001). Soft lithography in biology and biochemistry. *Annual Review of Biomedical Engineering*, **3**, 335–373.

[36] Wong, R. Y. M., Tso, C. Y., Fu, S. C., Chao, C. Y. H. (2022). Maxwell-Garnett permittivity optimized micro-porous PVDF/PMMA blend for near unity thermal emission through the atmospheric window. *Solar Energy Materials and Solar Cells*, **248**, 112003.

[37] Grum, F., Luckey, G. W. (1968). Optical sphere paint and a working standard of reflectance. *Applied Optics*, **7**, 2289–2294.

[38] De Sousa Meneses, D., Brun, J., Echegut, P., Simon, P. (2004). Contribution of semi-quantum dielectric function models to the analysis of infrared spectra. *Applied Spectroscopy*, **24**, 100658.

[39] Kelly-Gorham, M. R. K., DeVetter, B. M., Brauer, C. S., Cannon, B. D., Burton, S. D., Bliss, M., Johnson, T. J., Myers, T. L. (2017). Complex refractive index measurement for BaF_2 and CaF_2 via single-angle infrared reflectance spectroscopy. *Optical Materials*, **72**, 743–748.

[40] Palik, E. D. (1991). *Handbook of Optical Constants of Solids (Volume 2)*, Elsevier, United States of America.

[41] Best Research-Cell Efficiencies, https://www.nrel.gov/pv/assets/pdfs/best-research-cell-efficiencies.pdf, accessed in May 2023.

[42] Zhao, D., Yin, X., Xu, J., Tan, G., Yang, R. (2020). Radiative sky cooling-assisted thermoelectric cooling system for building applications. *Energy*, **190**, 116322.

[43] Kwan, T. H., Gao, D., Zhao, B., Ren, X., Hu, T., Dabwan, Y. N. (2021). Integration of radiative sky cooling to the photovoltaic and thermoelectric system for improved space cooling. *Applied Thermal Engineering*, **196**, 117230.

[44] Liu, Y., Hou, S., Wang, X., Yin, L., Wu, Z., Wang, X., Mao, J., Sui, J., Liu, X., Zhang, Q., Liu, Z., Cao, F. (2022). Passive radiative cooling enables improved performance in wearable thermoelectric generators. *Small*, **18**, 2106875.

[45] Khan, S., Kim, J., Roh, K., Park, G., Kim, W. (2021). High power density of radiative-cooled compact thermoelectric generator based on body heat harvesting. *Nano Energy*, **87**, 106180.

[46] Fan, L., Li, W., Jin, W., Orenstein, M., Fan, S. (2020). Maximum nighttime electrical power generation via optimal radiative cooling. *Optics Express*, **28**, 25460.

[47] Zhao, B., Pei, G., Raman, A. P. (2020). Modelling and optimization of radiative cooling based thermoelectric generators. *Applied Physics Letters*, **117**, 163903.

[48] Raman, A. P., Li, W., Fan, S. (2019). Generating light from darkness. *Joule*, **3**, 2679–2686.

[49] Ishii, S., Dao, T. D., Nagao, T. (2020). Radiative cooling for continuous thermoelectric power generation in day and night. *Applied Physics Letters*, **117**, 013901.

[50] Omair, Z., Assawaworrarit, S., Fan, L., Jin, W., Fan, S. (2022). Radiative-cooling based nighttime electricity generation with power density exceeding 100 mW/m^2. *iScience*, **25**, 104858.

[51] Ishii, S., Miura, A., Nagao, T., Uchida, K. (2021). Simultaneous harvesting of radiative cooling and solar heating for transverse thermoelectric generation. *Science and Technology of Advanced Materials*, **22**, 441–448.

[52] Li, M., Peterson, H. B., Coimbra, C. F. M. (2019). Radiative cooling resource maps for the contiguous United States. *Journal of Renewable and Sustainable Energy*, **11**, 036501.

[53] Zhu, Y., Qian, H., Yang, R., Zhao, D. (2021). Radiative sky cooling potential maps of China based on atmospheric spectral emissivity. *Solar Energy*, **218**, 195–210.

[54] Chen, J., Lu, L., Gong, Q. (2021). A new study on passive radiative sky cooling resource maps of China. *Energy Conversion and Management*, **237**, 114132.

[55] Huang, J., Lin, C., Li, Y., Huang, B. (2022). Effect of humidity, aerosol, and cloud on subambient radiative cooling. *International Journal of Heat and Mass Transfer*, **186**, 122438.

[56] Han, D., Fei, J., Li, H., Ng, B. F. (2022). The criteria to achieving sub-ambient radiative cooling and its limits in tropical daytime. *Building and Environment*, **221**, 109281.

[57] Bao, H., Yan, C., Wang, B., Fang, X., Zhao, C. Y., Ruan, X. (2017). Double-layer nanoparticle-based coatings for efficient terrestrial radiative cooling. *Solar Energy Materials and Solar Cells*, **168**, 78–84.

[58] Tso, C. Y., Chan, K. C., Chao, C. Y. H. (2017). A field investigation of passive radiative cooling under Hong Kong's climate. *Renewable Energy*, **106**, 52–61.

[59] Jeong, S. Y., Tso, C. Y., Ha, J. Y., Wong, Y. M., Chao, C. Y. H., Huang, B., Qiu, H. (2019). Field investigation of a photonic multi-layered TiO$_2$ passive radiative cooler in sub-tropical climate. *Renewable Energy*, **146**, 44–55.

[60] Jeong, S. Y., Tso, C. Y., Wong, Y. M., Chao, C. Y. H., Huang, B. (2019). Daytime passive radiative cooling by ultra emissive bio-inspired polymeric surface. *Solar Energy Materials and Solar Cells*, **206**, 110296.

[61] Han, D., Ng, B. F., Wan, M. P. (2020). Preliminary study of passive radiative cooling under Singapore's tropical climate. *Solar Energy Materials and Solar Cells*, **206**, 110270.

[62] Wong, R. Y. M., Tso, C. Y., Jeong, S. Y., Fu, S. C., Chao, C. Y. H. (2023). Critical sky temperatures for passive radiative cooling. *Renewable Energy*, **211**, 214–226.

[63] Berk, A., Conforti, P., Kennett, R., Perkins, T., Hawes, F., van den Bosch, J. (2014). MODTRAN 6: A major upgrade of the MODTRAN radiative transfer code. *2014 6th Workshop on Hyperspectral Image and Signal Processing: Evolution in Remote Sensing*, 1–4.

[64] Berk, A., Acharya, P. K., Bernstein, L. S., Anderson, G. P., Lewis, P., Chetwynd, J. H., Hoke, M. L. (2008). Band model method for modeling atmospheric propagation at arbitrarily fine spectral resolution. US patent no: 7433806.

[65] Brunt, D. (1932). Notes on radiation in the atmosphere I. *Quarterly Journal of the Royal Meteorological Society*, **58**, 389–420.

[66] Li, M., Jiang, Y., Coimbra, C. F. M. (2017). On the determination of atmospheric longwave irradiance under all-sky conditions. *Solar Energy*, **144**, 40–48.

[67] Liou, K. N. (2002). *An Introduction to Atmospheric Radiation*, 2nd edition. Elsevier, United States of America.

[68] Wallace, J. M., Hobb, P. V. (2006). *Atmospheric Science. An Introductory Survey*, 2nd edition. Elsevier, Canada.

[69] Meinel, A. B., Meinel, M. P. (1976). *Applied Solar Energy. An Introduction*. Addison-Wesley, United States of America.

[70] Liou, K. N., Yang, P. (2016). *Light Scattering by Ice Crystals, Fundamental and Applications*, Cambridge University Press, United Kingdom.

[71] Lin, C., Li, Y., Chi, C., Kwon, Y. S., Huang, J., Wu, Z., Zheng, J., Liu, G., Tso, C. Y., Chao, C. Y. H., Huang, B. (2022). A solution-processed inorganic emitter with high spectral selectivity for efficient subambient radiative cooling in hot humid climates. *Advanced Materials*, **34**, 2109350.

[72] Fei, J., Han, D., Ge, J., Wang, X., Koh, S. W., Gao, S., Sun, Z., Wan, M. P., Ng, B. D., Cai, L., Li, H. (2022). Switchable surface coating for bifunctional passive radiative cooling and solar heating. *Advanced Functional Materials*, **32**, 2203582.

CHAPTER 9

Transparent Radiative Cooling Materials and Their Application to Energy Harvesting

Kotaro Kajikawa

Tokyo Institute of Technology, Yokohama, Japan

Cooling an object usually requires energy, such as electricity. However, if it were possible to cool an object without using energy, not only could energy be saved, but energy could also be generated using a thermoelectric module, which is commercially available. Electromagnetic waves are radiated from an object according to its temperature. Thus energy-free cooling is possible if we can solve the problem of discarding the radiated electromagnetic waves. This chapter provides a basic explanation of radiative cooling (RC). It also shows transparent RC materials that can be used for cooling buildings and solar cells. Furthermore, an electrical device that generates energy using it is described.

9.1 INTRODUCTION

In recent years, research on optical technology for handling heat has attracted much attention. This is because these studies can contribute to solving environmental and energy problems. Among them, materials and devices using heat dissipation by blackbody radiation have been studied. In particular, the radiation and absorption of mid-infrared light are related to blackbody radiation at room temperature [1–3]. In this chapter, blackbody radiation is discussed first. Then, passive radiative

DOI: 10.1201/9781003409090-9

cooling (RC) [4–37] and transparent RC materials in the visible light region for daytime RC are introduced [38–48]. The transparent RC materials in the visible light region are particularly interesting research topics. Using these materials, RC can be achieved without changing the color or appearance of buildings and structures. It also enables us to realize cooling solar cells without a power source [38–42, 49, 50]. Cooling of solar cells leads to higher efficiency of power generation and is expected to be one of the solutions to energy problems.

This chapter also introduces non-RC materials and their applications. Most materials transparent in the visible wavelength range are oxides and polymers. These exhibit some degree of RC. In some cases, it is better to hinder RC. An example is the phenomenon of automobile windshields freezing on winter mornings. If RC can be reduced, it will be possible to realize transparent materials that do not radiate heat. Finally, as one of the applications of transparent RC materials, we describe an energy-harvesting device using such materials. This device can generate electricity day and night.

9.2 RADIATIVE COOLING

9.2.1 Blackbody Radiation

Objects emit electromagnetic waves (light), such as infrared and visible. This is called thermal radiation. Its spectrum varies with temperature T. The emissive intensity, $I(\lambda)$, of the light from a blackbody as a function of wavelength λ is

$$I(\lambda) = \frac{2hc^2}{\lambda^5} \frac{1}{\exp\left(\dfrac{hc}{\lambda kT}\right) - 1}, \tag{9.1}$$

where h is Planck's constant, k is Boltzmann's constant, and c is the velocity of light [1–3, 22]. A blackbody is a virtual object that absorbs light of all wavelengths. This is known as Planck's formula.

The relation between the light intensity and the emissive power, $E(\lambda)$, is given in the spherical coordinate with the axis of rotation normal to the blackbody surface as

$$E(\lambda) = \int_0^{2\pi} d\phi \int_0^{\pi/2} d\theta \ I(\lambda)\sin\theta\cos\theta = \pi I(\lambda), \tag{9.2}$$

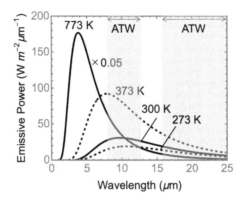

FIGURE 9.1   Blackbody Radiation at Different Temperatures.

where θ is the polar angle, and ϕ is the azimuthal angle [22]. The monochromatic emissivity spectra of a blackbody at several temperatures are shown in Figure 9.1. The peak of thermal radiation becomes a shorter wavelength with increasing temperature, and its intensity increases exponentially with increasing temperature.

The total emission power E_B corresponds to the radiation power from the blackbody, can be obtained by integrating Equation 9.2 with respect to the wavelength, and is given by

$$E_B = \sigma T^4. \tag{9.3}$$

The radiative power is proportional to the fourth power of the absolute temperature T, and the proportionality constant is $\sigma = 5.67 \times 10^{-8}$ W m^{-2} K^{-4}. For example, at $T = 300$ K, $E_B = 457$ Wm^{-2}. The wavelength range in Figure 9.1 is 0–25 μm, and the emission power of thermal radiation in this wavelength range is 381 Wm^{-2}, which is 83% of the total emission power. For a rigorous discussion, it is necessary to consider thermal radiation at wavelengths longer than 25 μm. Since it is difficult to make optical measurements in that wavelength range, in most cases, we will consider the wavelength range from 0.4 to 25 μm.

9.2.2 Atmospheric Transparency Window

Although the atmosphere surrounding the Earth appears transparent to our eyes, there is a wide range of wavelengths in the infrared region that is opaque due to the absorption of various gases, including carbon dioxide and water vapor. However, the wavelength regions from 8 to 13 μm

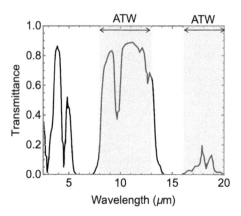

FIGURE 9.2   An example of atmospheric transparency in mid-latitude summer calculated by the radiative transfer model MODTRAN.

and from 16 to 25 µm are called the atmospheric transparency window (ATW), where the transmittance is high. Figure 9.2 shows an example of atmospheric transparency in mid-latitude summer calculated by the radiative transfer model MODTRAN [51]. Transmittance varies greatly with latitude and the amount of water vapor in the atmosphere. The gray region is the wavelength of the ATW. The wavelength region from 16 to 25 µm is sensitive to the amount of water vapor, and the transmittance is low under the condition that the water vapor content is high.

Interestingly, the peak of the monochromatic emissivity spectrum near room temperature ($T = 300$ K) is almost coincident with the ATW. Thus, heat from objects on the ground is radiated into space. This is the principle of RC. This phenomenon observed around us includes a drop in temperature on clear winter mornings when the wind is light. However, RC does not occur only in winter mornings but also in summer and even during the daytime. We do not notice it because of the strong sunlight and the fact that our own thermal radiation is in equilibrium with the thermal radiation from the surroundings, such as buildings, trees, walls, ceilings, etc. If we actively use RC, we can cool objects without using electrical energy.

The maximum amount of heat that can be radiated from the ground to space can be calculated by multiplying and integrating the total radiation capacity and the atmospheric permeability. The atmospheric transmissivity, shown in Figure 9.2, is 124 Wm^{-2} at $T = 300$ K. Most of this is due to thermal radiation through the ATW at short wavelengths (8–13 µm). In principle, a cooling capacity of 2 kW can be obtained with a 4 m-square heat radiator. This is comparable to the cooling capacity of a small air conditioner.

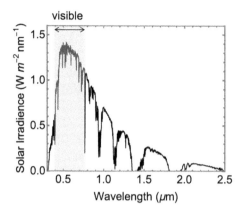

FIGURE 9.3   A sunlight spectrum (AM1.5).

9.2.3 Sunlight Spectrum

Figure 9.3 shows the sunlight spectrum. This is the spectrum AM1.5 (Airmass 1.5) used in the mid-latitude zone. The number after AM indicates the ratio of the path length of light through the atmosphere to the thickness of the atmosphere. For example, AM0 is the solar high spectrum outside the atmosphere. AM1.0 is the case when the sun is at the zenith (e.g., at the equatorial south of the equator at the vernal or autumnal equinoxes). AM1.5 is the case when the sun is at 42° altitude in the mid-latitude zone. In the mid-latitude zone, AM1.5 is often used as a representative value. Integrating this value by wavelength yields the energy of sunlight, which is about 900 Wm^{-2}. Since the energy falling from the sun to the earth is 1370 Wm^{-2} outside the atmosphere (solar constant), the energy is reduced to about 65% in the atmosphere. On the other hand, the amount of heat that can be radiatively cooled (127 W m^{-2}), calculated in the previous section, is 1/7 of this energy, indicating that the amount of heat that can be radiatively cooled is relatively large.

9.2.4 Emissivity

In Section 10.2.1, we discussed blackbody radiation, but blackbodies do not exist in reality. Therefore, it is necessary to consider the absorption coefficient A for a real object. The emissivity E, which is the ratio of thermal radiation of a real object to blackbody radiation, is equal to the absorption coefficient A (Kirchhoff's law) [1]. While it is difficult to measure emissivity, it is easy to measure absorption coefficient. Table 9.1 shows the average emissivity in the infrared region for various substances [1].

TABLE 9.1   Normal Total Emissivity $\bar{\varepsilon}$ and Normal Total Absorptivity $\bar{\alpha}$ for Incident Solar Radiation

Material		$\bar{\varepsilon}$	$\bar{\alpha}$
aluminum	polished	0.095	0.20
aluminum	foil	0.04	
gold	polished	0.018	0.29
silver	polished	0.01–0.03	
red brick		0.93	0.75
cotton	cloth	0.77	
ice	smooth	0.97	
paper	white	0.95	0.28
wood	oak	0.90	
paint	black	0.96–0.98	0.90
water		0.96	

The values are taken from Appendix B of Reference [1].

Materials show relatively high emissivity, but the high emissivity is over the infrared wavelength range. Thus the absorption at wavelengths other than the ATW is also large. This is not good for efficient RC. On the other hand, metals are among the materials with low emissivity. This is due to their high reflectance over the wide spectrum range.

9.2.5 RC Materials

In this section, we consider the properties of materials necessary to radiate heat into space efficiently. First, it is important to have a high emissivity at the wavelength of the ATW. In this wavelength range, there exists resonant absorption due to various molecular vibrations. In particular, strong vibrations due to carbon-fluorine bonds are observed at 7.8 to 10 μm (1000–1300 cm^{-1}). Absorption originating from Si-O resonance absorption in glass and other materials exists at 8.2–9.5 μm (1050–1220 cm^{-1}). [48, 52–59]. It is also known that many metal oxides have resonance in this wavelength range.

On the other hand, strong resonant absorption does not always give good results. Figure 9.4 shows the emissivity spectrum and refractive index of quartz glass [60]. Quartz shows a strong resonance around 9 μm, and the emissivity decreases significantly around this wavelength. This is because the real part of the refractive index becomes less than 1 due to the resonance of quartz, resulting in total reflection and an increase in the reflectance R. On a flat surface where scattering is negligible, the emissivity E is

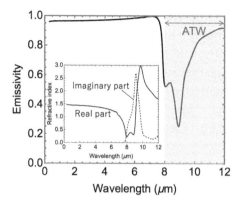

FIGURE 9.4   An emissivity spectrum of a quartz plate.

given by $A = 1 - R$. Therefore, E decreases as R increases. The same is true for glasses other than quartz (e.g., soda-lime glass). To prevent this, it is effective to roughen the glass surface or adsorb silica particles to reduce the reflectance. On the other hand, these measures reduce the transmittance, which is not a good way when transparent materials are required in the visible light range.

Subtracting the atmospheric transmittance from 1 gives the light absorption rate by the atmosphere. In the wavelength region outside the ATW, thermal radiation is also generated due to water vapor in the atmosphere, and its absorption raises the temperature of the object. This thermal radiation corresponding to the temperatures near the ground surface to the temperature above the sky is distributed over a wide wavelength range from 5 to 25 μm. Therefore, the light absorption in the wavelength range other than the ATW should be small, as described below. In order to achieve efficient RC, the material must have high emissivity in the wavelength band of the ATW and low emissivity, i.e., low absorption, in the other wavelength bands. Most of the materials listed in Table 9.1 do not meet the latter requirement, and therefore they do not make good RC materials.

In practice, they must have the optical properties required for each application. For nighttime RC, only the above-mentioned properties are required, but for daytime RC, various other requirements must be satisfied. For example, cooling solar cells requires high transmittance to all sunlight in addition to RC properties. Cooling of buildings and other structures requires materials that reflect or scatter sunlight to prevent its absorption and avoid temperature increase. In the case of clothing, color and texture are also important factors, so the quality of a material cannot be determined solely by its RC performance. In addition, for practical use, materials with low weather resistance and environmental load and low-cost materials are required.

Many materials have been reported for the purpose of daytime RC [38–48]. Some of them are based on metals, and others on metamaterials. Among them, materials with dispersed silica particles have been studied extensively, and some of them have been put into practical use. Some existing materials also show good properties. We have reported that acrylic polymers such as polymethyl methacrylate (PET) and polyethylene naphthalate (PEN) have good properties for RC [61]. Figure 9.5 shows their emissivity spectra. While they exhibit high emissivity (absorption) due to molecular vibration in the wavelength region of the ATW, they have relatively low emissivity outside the wavelength region of the ATW. PET is not easily affected by blackbody radiation from surrounding objects or the atmosphere. It is transparent in the visible light range and can be easily processed. It also has the advantages of relatively high weather resistance and low cost. Therefore, it is expected to be used as a daytime RC material.

9.2.6 Non-RC Materials

Most transparent materials, such as glass and acrylic resin, have absorption in the wavelength range of the ATW and thus exhibit a certain RC property. However, there are cases where this is a problem. For example, the windshield of a car freezes on a sunny winter morning because the

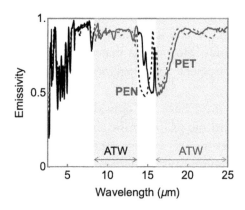

FIGURE 9.5 An emissivity spectra of PET and films.

temperature of the glass is lower than the air temperature. The car cannot be moved until the ice melts. To prevent this problem, windshields are covered with cloth, etc., but it would be convenient to solve this problem with materials and surface treatments that do not freeze.

Contrary to the previous policy, we have investigated for non-emissive cooling materials that have low emissivity in the wavelength region of the ATW and are transparent in the visible region. Metals have low emissivity but are not transparent. We prepared a film of silver nanowires dispersed on glass and confirmed that the temperature decrease due to RC could be suppressed. There is a trade-off between suppression of RC and transmittance in the visible light range. Therefore, experiments were conducted by changing the number of silver nanowires. It is difficult to match the experimental conditions, such as weather, temperature, wind speed, etc., outdoors. In order to make comparisons under well-defined conditions, we conducted radiation cooling experiments indoors using liquid nitrogen as a refrigerant. As a result, it was found that for a silver nanowire layer with a transmittance of 73%, the temperature drop could be suppressed to 5.2°C. On the other hand, the temperature drop of the glass was reduced to 9°C under the same conditions. On the other hand, the temperature drop of the glass was as large as 9.2°C under the same conditions. Although these non-RC performance is not yet satisfactory, we believe that higher performance non-RC materials can be obtained in the future by solidifying the silver nanowire layer with polymer resin or by changing the thickness of the nanowires.

9.3 APPLICATION OF RC TO ENERGY HARVESTING

9.3.1 Energy Harvesting

Energy harvesting is a technology to obtain electric power by utilizing energy in our environment, such as indoor light, vibration, radio waves, and heat. Although the power obtained is small, ranging from a few μW to mW, it has various advantages, such as no need for wiring for the power supply, no need to replace batteries, and a small and simple structure. Therefore, it has advantages that are not found in existing electric energy supply methods, including reduction of the environmental burden such as disposal of batteries, reduction of maintenance, weight reduction due to reduced wiring, and use in remote areas. It is expected to be used not only outdoors, where the grid power is absent, but also in IoT devices, wearable terminals, aircraft, drones, etc.

9.3.2 Energy Havesting Device Using RC

Solar cells can generate electricity during the daytime. Thus it is necessary to store electricity in storage batteries for nighttime. This requires the use of a storage site, which requires maintenance. Therefore, it makes sense to develop a simple system that can generate electricity at night.

Raman et al. reported about 0.75 mW of electric power obtained by using a thermoelectric module (TEM) to utilize the temperature difference generated by RC [62]. The TEM is attached to a disk-shaped radiator coated with black body paint. When it is exposed to the air, the temperature is lowered by RC. A metal fin is attached to the opposite side of the TEM, i.e., the hot part. The reason for this is that the TEM has heat conduction so that after a while, the temperature of the entire element becomes the same, and no electromotive force is generated. When normalized to the area of a 200 mm diameter radiator, the obtained power was 25 mWm⁻².

In addition to power generation by RC, Munday et al. also mentioned the possibility of nighttime power generation using pn junctions [63]. The thermally excited thermal radiation produced by a pn junction with a small band gap is emitted into space. This is thermoradiative photovoltaics and has been the subject of several theoretical studies. Experimental studies of thermal photovoltaics have also been reported, but no significant results have been obtained.

9.3.3 Improvement of Energy Harvesting Devices

In order to improve the performance of the power generation using RC, we have fabricated a multi-stage TEM, as shown in Figure 9.6, using a thermal storage unit with water instead of fins. The multiple-stage TEM is expected to have the effect of securing the temperature difference between the high and low-temperature sides by increasing the thermal

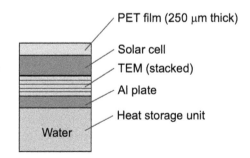

FIGURE 9.6 A schematic of the energy harvesting device.

TABLE 9.2   Summery of Generated Power (Wm^{-2})
Measured in the Experiments

	Solar Cell	TEM
Daytime	71	3.1
Nighttime	–	0.069

resistance of the elements as a whole. Furthermore, in order to generate electricity using solar cells during the daytime, solar cells are placed
between the PET film used as a radiator and the TEM. Since the PET is
transparent, it does not affect the operation of the solar cells. Therefore,
the device can generate electricity day and night. First, the relationship
between the number of multi-staged TEMs and the obtained power was
examined by measuring the electromotive force indoors using liquid
nitrogen as a refrigerant to determine the optimal number of TEMs. This
is due to the increase of internal resistance as the number of TEMs is
increased.

The temperature of the PET film used as a radiator decreased, and electromotive force (EMF) was generated. On the other hand, the heat sink is
operating as expected with almost constant temperature. The EMF measured by the voltmeter is the open circuit voltage. In order to estimate the
available energy, it is essential to measure the power. Therefore, the current-
voltage characteristic was measured using a source measure unit (SMU). As
a result, 0.19 mW of power was obtained from the TEM. When this power
is normalized by the area of PET used as a radiator, 69 mWm^{-2} is obtained,
which is 2.8 times larger than the value reported by Raman et al. [62].

During the daytime, the solar cells can generate electricity from sunlight, and the TEM can also generate electricity using the heat generated by
sunlight. The results of these measurements are summarized in Table 9.2.
Amorphous solar cells are used here. It is natural that the power generated
during the daytime is larger than that obtained during the nighttime, and
the power generated during the daytime by TEM is relatively high, about
5% of that by solar cells. The challenge for the future is to increase the
power generated during the nighttime through further improvements.

9.4 SUMMARY

In this chapter, the principle of RC is extended, and transparent RC materials are reviewed. We also discuss materials that are transparent but have
low RC. In addition, our attempts to develop an energy harvesting device

are presented. We obtained 2.8 times higher power than previous studies. Since RC can contribute to solving energy problems, future research progress is expected.

BIBLIOGRAPHY

[1] J. R. Howell, M. P. Mengüç, Daun K., and R Siegel. *Thermal Radiation Heat Transfer* 7th Edition. CRC Press, 2021.

[2] B. E. A. Saleh and M. C. Teich. *Fundamentals of Photonics*. Wilely Interscience, 2007.

[3] G. R. Fowles. *Introduction to Modern Optics* 2nd Edition. Dover, 1975.

[4] B. Zhao, M. Hu, X. Ao, N. Chen, and G. Pei. Radiative cooling: A review of fundamentals, materials, applications, and prospects. *Appl. Energy*, Vol. 236, pp. 489–513, 2019.

[5] M. M. Hossain and M. Gu. Radiative cooling: Principles, progress, and potentials. *Adv. Sci.*, Vol. 3, No. 7, p. 1500360, 2016.

[6] M. Zeyghami, D. Y. Goswami, and E. Stefanakos. A review of clear sky radiative cooling developments and applications in renewable power systems and passive building cooling. *Sol. Energy Mater. Sol. Cells*, Vol. 178, pp. 115–128, 2018.

[7] X. Sun, Y. Sun, Z. Zhou, M. A. Alam, and P. Bermel. Radiative sky cooling: fundamental physics, materials, structures, and applications. *Nanophotonics*, Vol. 6, No. 5, pp. 997–1015, 2017.

[8] D. Zhao, A. Aili, Y. Zhai, S. Xu, G. Tan, X. Yin, and R. Yang. Radiative sky cooling: Fundamental principles, materials, and applications. *Appl. Phy. Rev.*, Vol. 6, No. 2, p. 021306, 2019.

[9] S. Vall and A. Castell. Radiative cooling as low-grade energy source: A literature review. *Renew. Sustain. Energy Rev.*, Vol. 77, pp. 803–820, 2017.

[10] J. P. Bijarniya, J. Sarkar, and P. Maiti. Review on passive daytime radiative cooling: Fundamentals, recent researches, challenges and opportunities. *Renew. Sustain. Energy Rev.*, Vol. 133, p. 110263, 2020.

[11] A. S. Farooq, P. Zhang, Y. Gao, and R. Gulfam. Emerging radiative materials and prospective applications of radiative sky cooling – A review. *Renew. Sustain. Energy Rev.*, Vol. 144, p. 110910, 2021.

[12] J. Zhang, J. Yuan, J. Liu, Z. Zhou, J. Sui, J. Xing, and J. Zuo. Cover shields for sub-ambient radiative cooling: A literature review. *Renew. Sustain. Energy Rev.*, Vol. 143, p. 110959, 2021.

[13] K.-T. Lin, J. Han, K. Li, C. Guo, H. Lin, and B. Jia. Radiative cooling: Fundamental physics, atmospheric influences, materials and structural engineering, applications and beyond. *Nano Energy*, Vol. 80, p. 105517, 2021.

[14] B. Ko, D. Lee, T. Badloe, and J. Rho. Metamaterial-based radiative cooling: Towards energy-free all-day cooling. *Energies*, Vol. 12, No. 1, p. 89, 2018.

[15] J. Chen and L. Lu. Development of radiative cooling and its integration with buildings: A comprehensive review. *Sol. Energy*, Vol. 212, pp. 125–151, 2020.

[16] X. Yu, J. Chan, and C. Chen. Review of radiative cooling materials: Performance evaluation and design approaches. *Nano Energy*, Vol. 88, p. 106259, 2021.

[17] M. Chen, D. Pang, X. Chen, H. Yan, and Y. Yang. Passive daytime radiative cooling: Fundamentals, material designs, and applications. *EcoMat.*, Vol. 4, No. 1, p. e12153, 2021.

[18] Z. Li, Q. Chen, Y. Song, B. Zhu, and J. Zhu. Fundamentals, materials, and applications for daytime radiative cooling. *Adv. Mater. Technol.*, Vol. 5, No. 5, p. 1901007, 2020.

[19] J. Liu, Z. Zhou, J. Zhang, W. Feng, and J. Zuo. Advances and challenges in commercializing radiative cooling. *Mater. Today Phys.*, Vol. 11, p. 100161, 2019.

[20] X. Yin, R. Yang, G. Tan, and S. Fan. Terrestrial radiative cooling: Using the cold universe as a renewable and sustainable energy source. *Science*, Vol. 370, pp. 786–791, 2020.

[21] Y. Yang and Y. Zhang. Passive daytime radiative cooling: Principle, application, and economic analysis. *MRS Energy Sustain.*, Vol. 7, No. 1, p. e18, 2020.

[22] A. P. Raman, M. A. Anoma, L. Zhu, E. Rephaeli, and S. Fan. Passive radiative cooling below ambient air temperature under direct sunlight. *Nature*, Vol. 515, No. 7528, pp. 540–544, 2014.

[23] Y. Zhai, Y. Ma, S. N. David, D. Zhao, R. Lou, G. Tan, R. Yang, and X. Yin. Scalable-manufactured randomized glass-polymer hybrid metamaterial for daytime radiative cooling. *Science*, Vol. 355, pp. 1062–1066, 2017.

[24] M. Qi, T. Wu, Z. Wang, Z. Wang, B. He, H. Zhang, Y. Liu, J. Xin, T. Zhou, X. Zhou, and L. Wei. Progress in metafibers for sustainable radiative cooling and prospects of achieving thermally drawn metafibers. *Adv. Energy Sustainability Res.*, 2021.

[25] D. Sato and N. Yamada. Review of photovoltaic module cooling methods and performance evaluation of the radiative cooling method. *Renew. Sustain. Energy Rev.*, Vol. 104, pp. 151–166, 2019.

[26] T. Suichi, A. Ishikawa, Y. Hayashi, and K. Tsuruta. Performance limit of daytime radiative cooling in warm humid environment. *AIP Adv.*, Vol. 8, No. 5, 2018.

[27] Y. Li, L. Li, L. Guo, and B. An. Systematical analysis of ideal absorptivity for passive radiative cooling. *Opt. Mater. Express*, Vol. 10, No. 8, 2020.

[28] P. Li, A. Wang, J. Fan, Q. Kang, P. Jiang, H. Bao, and X. Huang. Thermo-optically designed scalable photonic films with high thermal conductivity for subambient and above-ambient radiative cooling. *Adv. Funct. Mater.*, p. 2109542, 2021.

[29] L. Zhou, H. Song, J. Liang, M. Singer, M. Zhou, E. Stegenburgs, N. Zhang, C. Xu, T. Ng, Z. Yu, and B. Ooi, and Q. Gan. A polydimethylsiloxane-coated metal structure for all-day radiative cooling. *Nat. Sustain.*, Vol. 2, pp. 718–724, 2019.

[30] J.-L. Kou, Z. Jurado, Z. Chen, S. Fan, and A. J. Minnich. Daytime radiative cooling using near-black infrared emitters. *ACS Photonics*, Vol. 4, No. 3, pp. 626–630, 2017.

[31] P.-C. Hsu, A. Y. Song, P. B. Catrysse, C. Liu, Y. Peng, J. Xie, S. Fan, and Y. Cui. Radiative human body cooling by nanoporous polyethylene textile. *Science*, Vol. 353, pp. 1019–1023, 2016.

[32] Y. Sun, Y. Ji, M. Javed, X. Li, Z. Fan, Y. Wang, Z. Cai, and B. Xu. Preparation of passive daytime cooling fabric with the synergistic effect of radiative cooling and evaporative cooling. *Adv. Mat. Tech.*, p. 2100803, 2021.

[33] M. G. Abebe, G. Rosolen, E. Khousakoun, J. Odent, J.-M. Raquez, S. Desprez, and B. Maes. Dynamic thermal-regulating textiles with metallic fibers based on a switchable transmittance. *Phys. Rev. App.*, Vol. 14, No. 4, p. 044030, 2020.

[34] X. Wang, X. Liu, Z. Li, H. Zhang, Z. Yang, H. Zhou, and T. Fan. Scalable flexible hybrid membranes with photonic structures for daytime radiative cooling. *Adv. Func. Mat.*, Vol. 30, No. 5, p. 1907562, 2019.

[35] L. Cai, A. Y. Song, W. Li, P. C. Hsu, D. Lin, P. B. Catrysse, Y. Liu, Y. Peng, J. Chen, H. Wang, J. Xu, A. Yang, S. Fan, and Y. Cui. Spectrally selective nano-composite textile for outdoor personal cooling. *Adv. Mater.*, Vol. 30, No. 35, p. e1802152, 2018.

[36] S. Zeng, S. Pian, M. Su, Z. Wang, M. Wu, X. Liu, M. Chen, Y. Xiang, J. Wu, M. Zhang, Q. Cen, Y. Tang, X. Zhou, Z. Huang, R. Wang, A. Tunuhe, X. Sun, Z. Xia, M. Tian, M. Chen, X. Ma, L. Yang, J. Zhou, H. Zhou, Q. Yang, X. Li, Y. Ma, and G. Tao. Hierarchical-morphology metafabric for scalable passive daytime radiative cooling. *Science*, Vol. 373, pp. 692–696, 2021.

[37] S. Gamage, E. S. H. Kang, C. Akerlind, S. Sardar, J. Edberg, H. Kariis, T. Ederth, M. Berggren, and M. P. Jonsson. Transparent nanocellulose meta-material enables controlled optical diffusion and radiative cooling. *J. Mat. Chem. C*, Vol. 8, No. 34, pp. 11687–11694, 2020.

[38] Z. Li, S. Ahmed, and T. Ma. Investigating the effect of radiative cooling on the operating temperature of photovoltaic modules. *Solar RRL*, Vol. 5, p. 2000735, 2021.

[39] K. Won Lee, W. Lim, M. S. Jeon, H. Jang, J. Hwang, C. H. Lee, and D. R. Kim. Visibly clear radiative cooling metamaterials for enhanced thermal management in solar cells and windows. *Adv. Funct. Mater.*, p. 2105882, 2021.

[40] W. Li, Y. Shi, K. Chen, L. Zhu, and S. Fan. A comprehensive photonic approach for solar cell cooling. *ACS Photonics*, Vol. 4, No. 4, pp. 774–782, 2017.

[41] E. Lee and T. Luo. Black body-like radiative cooling for flexible thin-film solar cells. *Sol. Energy Mater. Sol. Cells*, Vol. 194, pp. 222–228, 2019.

[42] G. Chen, Y. Wang, J. Qiu, J. Cao, Y. Zou, S. Wang, J. Ouyang, D. Jia, and Y. Zhou. A visibly transparent radiative cooling film with self-cleaning function produced by solution processing. *J. Mat. Sci. Technol.*, Vol. 90, pp. 76–84, 2021.

[43] D. Fan, H. Sun, and Q. Li. Thermal control properties of radiative cooling foil based on transparent fluorinated polyimide. *Sol. Energy Mater. Sol. Cells*, Vol. 195, pp. 250–257, 2019.

[44] Y. Tian, L. Qian, X. Liu, A. Ghanekar, G. Xiao, and Y. Zheng. Highly effective photon-to-cooling thermal device. *Sci. Rep.*, Vol. 9, p. 19317, 2019.

[45] D. Li, X. Liu, W. Li, Z. Lin, B. Zhu, Z. Li, J. Li, B. Li, S. Fan, J. Xie, and J. Zhu. Scalable and hierarchically designed polymer film as a selective thermal emitter for high-performance all-day radiative cooling. *Nat. Nanotechnol.*, Vol. 16, No. 2, pp. 153–158, 2021.

[46] K. Wang, G. Luo, X. Guo, S. Li, Z. Liu, and C. Yang. Radiative cooling of commercial silicon solar cells using a pyramid-textured pdms film. *Solar Energy*, Vol. 225, pp. 245–251, 2021.

[47] M.-Q. Lei, Y.-F. Hu, Y.-N. Song, Y. Li, Y. Deng, K. Liu, L. Xie, J.-H. Tang, D.-L. Han, J. Lei, and Z.-M. Li. Transparent radiative cooling films containing poly(methylmethacrylate), silica, and silver. *Opt. Mat.*, Vol. 122, p. 111651, 2021.

[48] A. Aili, Z. Y. Wei, Y. Z. Chen, D. L. Zhao, R. G. Yang, and X. B. Yin. Selection of polymers with functional groups for daytime radiative cooling. *Mat. Today Phys.*, Vol. 10, p. 100127, 2019.

[49] L. Zhu, A. Raman, K. X. Wang, M. A. Anoma, and S. Fan. Radiative cooling of solar cells. *Optica*, Vol. 1, No. 1, pp. 32–38, 2014.

[50] Z. Wang, D. Kortge, J. Zhu, Z. Zhou, H. Torsina, C. Lee, and P. Bermel. Lightweight, passive radiative cooling to enhance concentrating photovoltaics. *Joule*, Vol. 4, No. 12, pp. 2702–2717, 2020.

[51] MODTRAN. Website. http://modtran.spectral.com/modtran_home

[52] S. Meng, L. Long, Z. Wu, N. Denisuk, Y. Yang, L. Wang, F. Cao, and Y. Zhu. Scalable dual-layer film with broadband infrared emission for sub-ambient daytime radiative cooling. *Sol. Energy Mater. Sol. Cells*, Vol. 208, p. 110393, 2020.

[53] P. Yang, C. Chen, and Z. M. Zhang. A dual-layer structure with record-high solar reflectance for daytime radiative cooling. *Solar Energy*, Vol. 169, pp. 316–324, 2018.

[54] J. Mandal, Y. Fu, A. C. Overvig, M. Jia, K. Sun, N. N. Shi, H. Zhou, X. Xiao, N. Yu, and Y. Yang. Hierarchically porous polymer coatings for highly efficient passive daytime radiative cooling. *Science*, Vol. 362, pp. 315–319, 2018.

[55] H. Zhang and D. Fan. Improving heat dissipation and temperature uniformity in radiative cooling coating. *Energy Technol.*, Vol. 8, No. 5, p. 1901362, 2020.

[56] W. Huang, Y. Chen, Y. Luo, J. Mandal, W. Li, M. Chen, C.-C. Tsai, Z. Shan, N. Yu, and Y. Yang. Scalable aqueous processing-based passive daytime radiative cooling coatings. *Adv. Funct. Mater.*, Vol. 31, p. 2010334, 2021.

[57] J. Liu, J. Zhang, D. Zhang, S. Jiao, Z. Zhou, H. Tang, J. Zuo, and Z. Zhang. Theoretical and experimental research towards the actual application of sub-ambient radiative cooling. *Sol. Energy Mater. Sol. Cells*, Vol. 220, p. 110826, 2021.

[58] H. Zhong, P. Zhang, Y. Li, X. Yang, Y. Zhao, and Z. Wang. Highly solar-reflective structures for daytime radiative cooling under high humidity. *ACS Appl. Mater. Interfaces*, Vol. 12, No. 46, pp. 51409–51417, 2020.

[59] H. T. T. Tam, M. Toma, T. Okamoto, Hidaka M., K. Fujii, Y. Kuwana, and K. Kajikawa. Weatherable, solvent-soluble, paintable and transparent fluoropolymers for daytime radiative cooling. *Int. J. Therm. Sci.*, Vol. 184, p. 107959, 2023.

[60] R. Kitamura, L. Pilon, and M. Jonasz. Optical constants of silica glass from extreme ultraviolet to far infrared at near room temperature. *Appl. Opt.*, Vol. 46, pp. 8118–8133, 2007.

[61] H. T. T. Tam, M. Toma, and K. Kajikawa. Investigation of polyesters as daytime radiative cooling materials. *Mol. Cryst. Liq. Cryst.*, Vol. 741, pp. 17–23, 2022.

[62] A. P. Raman, W. Li, and S. Fan. Generating light from darkness. *Joule*, Vol. 3, pp. 2679–2686, 2019.

[63] T. Deppe and J. N. Munday. Nighttime photovoltaic cells: Electrical power generation by optically coupling with deep space. *ACS Photonics*, Vol. 7, No. 1, pp. 1–9, 2019.

CHAPTER 10

Metamaterials for High Efficiency Solar Desalination with Carbon Zero Emission

Fei Xiang and Andrea Fratalocchi

Primalight Group, Faculty of Electrical and Computer Engineering, King Abdullah University of Science and Technology (KAUST), Saudi Arabia

10.1 INTRODUCTION

Carbon-neutral or carbon-negative technologies attract considerable research interest in renewable energy and environmental protection to address the challenges of climate change [1, 2]. These technologies could help implement a circular carbon society with sustainable development while offering promising avenues for recycling environmental CO_2 into value-added products [3, 4].

Significant research efforts are devoted to advancing circular carbon technologies across diverse fields, including direct air capture [5], bioenergy production [6], and CO_2 reduction [7]. These areas focus on studying different mechanisms to capture emitted CO_2 and harness it for producing chemicals, fuels, and food [4, 7]. Other examples are carbon-neutral technologies, such as solar desalination. The scarcity of clean water is one of the most critical and pressing challenges confronting the development of modern society [8, 9]. Approximately one-fifth of the global population, comprising billions of people, resides in regions facing water scarcity problems [10].

DOI: 10.1201/9781003409090-10

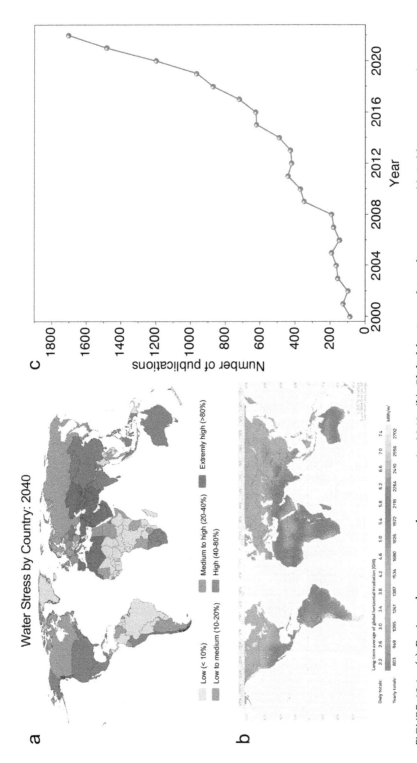

FIGURE 10.1 (a) Projected water stress by country in 2040. (b) Global horizontal irradiation map. (c) Publications on solar water desalination in the past 20 years. ((a) Adapted from [11], with permission. (b) Reproduced from Solargis [12], with permission.)

Moreover, an additional 1.6 billion individuals face significant constraints in accessing freshwater due to technical or financial limitations, despite available water resources in their proximity [8, 10, 11] (Figure 10.1a). The United Nations' World Water Development Report 2012 estimates that water scarcity will rapidly escalate to two-thirds of the global population by 2025 [10]. Addressing this problem offers a significant opportunity for research in an interdisciplinary area combining physics, engineering, chemistry, computer, and material sciences.

Because solar energy is the most abundant and renewable energy resource currently available on the planet [13, 14] (Figure 10.1b), solar water desalination gathers increasing attention as the next technology for clean and sustainable production of clean water from undrinkable water sources, such as seawater, lake/river water, and industrial wastewater [8, 15, 16]. Solar evaporation technologies yield high-purity freshwater with no additional chemicals and carbon-zero emissions, serving as an environmentally friendly approach to minimize the carbon footprint in the society [17, 18]. Over the past two decades, the scientific research in this area increased exponentially, with nearly 2000 articles in 2022 (Figure 10.1c). Achieving high-efficiency solar desalination relies on the design and engineering of solar evaporators [8, 15]. Extensive research efforts aim to explore and enhance the performance of solar evaporation systems [19–21], improving the overall solar-to-steam energy efficiency [22–24], and long-term operational stability [25, 26]. This chapter outlines recent advances in solar evaporator technology, focusing on materials engineering, interface design, and structural optimizations. It summarizes recent technological development explored using metamaterials, representing a class of artificially engineered complex structures that harness light energy controllably for various applications. This chapter discusses recent advances in the field, including a perspective on the research challenges, opportunities, and future directions in this area of research.

10.2 SOLAR WATER DESALINATION TECHNIQUES

10.2.1 Photo-Thermal Materials Engineering

Solar water desalination devices require specific materials to perform different functions, such as photothermal materials for light absorption, thermal-insulating materials for efficient thermal management, and water-transporting materials. Among them, converting solar energy into heat by photothermal materials is crucial in initiating solar vapor generation [16]. Recent developments in solar absorber materials have identified various

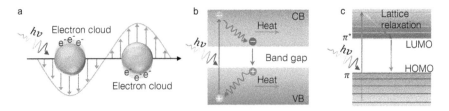

FIGURE 10.2 Typical types of photothermal mechanisms. (a) Plasmonic heating in plasmonic metal nanoparticles. (b) Electron-hole generation and relaxation in semiconductors. (c) Molecular heating.

promising candidates for photothermal conversion, including plasmonic metals [18, 22, 24, 27], semiconductor particles [28, 29], polymers [26, 30], and carbon-based nanostructures [9, 21, 25, 31]. The mechanisms investigated for photothermal conversion range from plasmonic localized heating in plasmonic metals (Figure 10.2a) to electron-hole generation and relaxation in semiconductors (Figure 10.2b), as well as thermal vibration of molecules in polymers and carbon-based materials (Figure 10.2c). This chapter explores diverse types of photothermal materials and their corresponding mechanisms and highlights recent studies conducted in each case.

10.2.1.1 Thermo-plasmonics

Metal nanoparticles, such as gold (Au), silver (Ag), and aluminum (Al), exhibit strong light absorption at specific frequencies and can convert light into heat via plasmon resonance [32, 33]. Plasmon resonance occurs when the frequency of incident light matches the natural oscillation frequency of the free electrons on the metal surface [8]. These electrons promote their energy from the Fermi level to higher energy levels and become "hot" during the energetic transition. Hot electrons can oscillate coherently with an incident electromagnetic field, converting absorbed light energy into heat through electron-phonon scattering. The generated heat eventually dissipates into the surrounding environment via phonon-phonon relaxation [16] (Figure 10.2a).

In recent years, many groups have investigated such excitation/relaxation processes in solar water desalination [34, 35]. When applying conventional metals to large-scale implementations of solar desalination, two main issues emerge. The first problem is surface plasmon resonances trigger from a narrow frequency range of incident light. This mechanism needs to broaden to enhance performance in solar-driven thermoplasmonics [32, 35]. The second issue is noble metals' cost and material scarcity [27, 36]. The work of Zhou *et al.* [24] explored the use of aluminum (Al) to address these problems. While Al is less expensive (US$1,000/ton) than gold or other noble metals, the Al plasmonic responses exist at ultraviolet frequencies [37, 38]. The authors of [24] implement a nanostructuring process that turns bare Al into a broadband plasmon-enhanced solar evaporator. The idea is to engineer a porous membrane from three-dimensional self-assembled Al nanoparticles (NPs). The authors use anodic oxidation of Al foil into a nanoporous aluminum oxide membrane (AAM), followed by physical vapor deposition (PVD). This process forms a thin Al layer on the top surface of AAM and close–packed aluminum nanoparticles along the sidewalls (Figure 10.3a). After Al deposition, the color of AAM changes from transparent to black, with a solar absorption of over 96% (Figure 10.3b, c).

The authors interpret the observed broadband solar absorption with finite difference time domain (FDTD) simulations. Computation models attribute the enhanced absorption performance to a synergistic effect. At first, the closely packed Al NPs on the sidewalls trigger strong plasmon hybridization [39] and excitation of high-density localized surface plasmons (LSP) resonances [40]. On the other hand, the concurring dielectric changes in the natural oxidation of Al NPs broaden the absorption bandwidth to the IR range, thus enhancing the overall absorption. Figure 10.3e provides simulation details of this mechanism by reporting results on a 2 nm thick Al oxide layer combined with a close-packed Al NPs geometry. Theoretically computed absorption shows good agreement with experimental results (Figure 10.3f).

In solar desalination experiments, Al NP/AAM plasmonic structures leverage localized heating from non-radiative plasmon decay (Figure 10.3g), with enhanced performances compared with pure water evaporation showing an evaporation rate at optical concentration $Copt = 4$ suns reaching ~5.7 kg m^{-2} h^{-1} and ~1.0 kg m^{-2} h^{-1} under one sun illumination. The steam generation efficiency reaches up to 88.4% and 91% under four sun and 6 sun illuminations, respectively. The authors report that this Al NP/AAM plasmonic evaporator operates in stable conditions for 20 cycles, with each cycle of 1-hour length (Figure 10.3k).

Another approach to overcoming the limits of narrow resonance frequencies in classical plasmonic materials is employing plasmonic metasurfaces. Liu *et al.* [22, 41] report a flat-optics metasurface, behaving as an almost ideal blackbody, engineered from suitably assembled Au nanoparticles with a wet-chemistry method. This dark metasurface exhibits complete light absorption at all frequencies and polarizations (averaged light

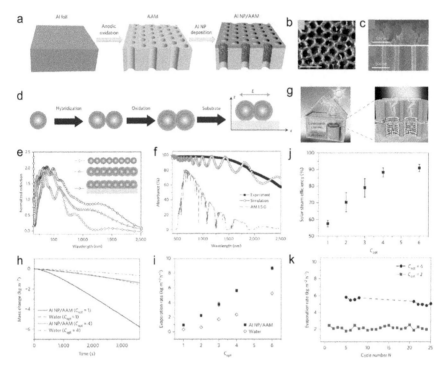

FIGURE 10.3 3D aluminium nanoparticles for plasmon-enhanced solar desalination. (a) Fabrication process of 3D Al-based plasmonic structures. (b) Top-view and (c) cross-sectional view of SEM images of the 3D Al-based plasmonic structure. (d) Absorption mechanisms of Al-based plasmonic structures. (e) Dependence of calculated normalized extinction cross-section (Cext/Cgeo) of Al NPs on the substrate. (f) Experimental and simulated absorption of aluminum-based plasmonic absorbers (inset dashed line: normalized spectral solar irradiance density of air mass 1.5 global (AM 1.5 G) tilt solar spectrum). (g) Experimental set-up for solar desalination. (h) Mass changes result with/without the Al NP/AAM structure under 1 and 4 sun irradiation. (i) The dependence of the evaporation rate of the Al NP/AAM structure and pure water on Copt (with dark evaporation subtracted). (j) Solar steam efficiency of the Al NP/AAM structure under different Copt. (k) Cycling performance of an Al NP/AAM structure under constant illuminations of 2 suns and four suns (1 h for each cycle). (Reproduced from Zhou et al. [24], with permission.)

absorption of 97%), has large-scale manufacturing, and is 98% recyclable. The authors fabricated the Au plasmonic nanostructures and dispersed these Au nanostructures on filter paper to construct blackbody metasurfaces (Figure 10.4a). The Au nanostructures consist of a nanorod (75 nm in length and 18 nm in diameter) combined with a nanosphere. The authors performed quantitative analysis on the conversion of electromagnetic

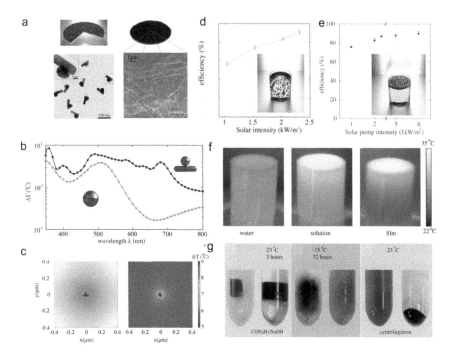

FIGURE 10.4 Super-dark flat-optics metasurface with Au nanoparticles. (a) 3D sketch of the sample film and SEM images of the Au nanoparticles and film. (b) Spectral dependence of the temperature increases $\Delta T(\lambda)$ for a blackbody nanostructure composing the flat-optics metasurface (circle markers) and a nanosphere (triangle markers). (c) Comparison of cross-sectional temperature increases ΔT at $\lambda = 510$ nm. (d) Thermal efficiency η as a function of solar intensity (I) for nanoparticles at a constant concentration in water ($C = 0.8$ g L^{-1}). (e) Solar thermal efficiency η as a function of the input intensity (inset: metasurface-based steam generation system). (f) Thermal images of different steam generation systems at steady state. (g) Demonstration of the recyclability of the nanoparticles. (Reproduced from Liu *et al.* [22], with permission.)

energy into heat with the finite-element method (FEM), observing a broadband temperature increase of $\overline{\Delta T} = 3.41$ °C (Figure 10.4b). This increase corresponds to a 153% enhancement compared with classical nanosphere (1.35 °C). When illuminated with light at the wavelength of $\lambda = 510$ nm, the flat-optic nanostructures show a wide region of spatial temperature increase. By comparison, classical Au nanospheres exhibit localized temperature variation within the structure (Figure 10.4c).

If applied to solar steam generation without any substrate, the black nanoparticles show a solar conversion efficiency of 65.5% at the input intensity of 2.3 kW m^{-2} with a nanoparticles concentration of 0.8 g L^{-1}

(Figure 10.4d). When combined with a paper substrate, the black metasurface (~200 nm) achieves a solar-to-heat conversion efficiency of 76% and 87% under one sun and 2.3 suns, respectively, with a water production rate of 1.18 kg m^{-2} h^{-1} at one sun (Figure 10.4e). The enhanced solar steam generation efficiency originates from a strongly induced heat localization on the top surface with a temperature difference up to 13 °C higher than the bulk water underneath (Figure 10.4f). The authors have demonstrated that these nanostructures are recyclable up to 98% (Figure 10.4g).

Constructing hierarchical nanostructures is another approach to achieving broadband solar absorption in plasmonic metals. Recently, Zhang *et al.* [18] report a controlled dealloying process on a dilute $Cu_{99}Au_1$ precursor and convert the precursor to self-supporting black gold (Au) film with hierarchical porous structures for solar steam generation (Figure 10.5a). The obtained nanoporous gold (NPG) exhibits a large volume contraction of ≈87% and ultrahigh prosperity above 80% (Figure 10.5b). The NPG film is lightweight due to its high porosity and low density (Figure 10.5c). It possesses super-hydrophilic properties, forming a zero-contact angle with water droplets within 6 s (Figure 10.5d). Benefiting from such a nanoporous structure, the NPG film with 6 h dealloying time (NPG-6) presents broadband light absorption from 300 nm to 2500 nm with an absorbance of 75.0–91.4% (Figure 10.5e). The authors implement a solar evaporation device combining NPG film with polystyrene foam as a thermal insulator and cotton for water transport (Figure 10.5f). Infrared images demonstrate the rapid photothermal response in NPG film and localized heat distribution in the surface area (Figure 10.5g). The NPG-6 shows larger mass changes of water than NPG-30 under light intensities of 1, 3, and 5 kW m^{-2} (Figure 10.5h) and reports a photothermal conversion efficiency up to 95.5% and an evaporation rate of 1.51 kg m^{-2} h^{-1} under one sun (Figure 10.5i). Moreover, this NPG-based solar evaporator maintains an average evaporation rate of 1.50 kg m^{-2} h^{-1} over 18 cycles in continuous solar steam generation measurements (Figure 10.5j).

The authors attribute the performances of this NPG evaporator to a combination of synergistic effects: broadband light absorption, enhanced water transport paths of micro/nanochannels in the film, and reduced heat dissipation in the thermal insulation structure design. The results yield a solar energy conversion efficiency of 94.5% under one sun.

Besides solar steam generation performances, fabrication costs of evaporators are a crucial factor for the industrial scalability of this technology in the future [9]. Carbon materials are a promising candidate to address this

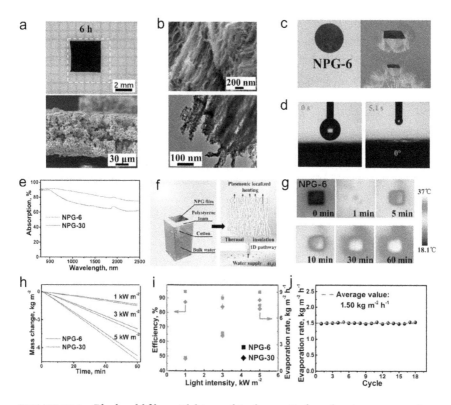

FIGURE 10.5 Black gold film with hierarchical porosity for solar steam generation. (a-b) Real-size photograph, SEM, and TEM images of the black Au porous film. (c) Image of fabricated samples, (d) water contact angle measurements, and (e) black Au porous film absorption spectra. (f) Schematic of the experimental setup for solar steam generation. (g) Infrared images of the black Au film under a light intensity of 1 Sun. (h) Water mass change under different light intensities. (i) Photothermal conversion efficiency and evaporation rate of the NPG-6 and NPG-30 films at different light intensities. (j) Cyclic water evaporation test of the NPG-6 film. (Reproduced from Zhang *et al.* [18], with permission.)

problem due to their low cost and abundant availability in nature [42, 43]. Besides that, developing new uses of carbon in carbon neutral or carbon negative technologies to mitigate global warming effects is a major sustainable development goal (SDG) [44]. Mazzone *et al.* [9] implements a solar steam generator by engineering the porosity of natural coal and combining it with cotton fibers. The authors have manufactured a carbonized compressed powder (CCP) from coconut shells and mixed it with cotton fibers. In the final device, the cotton fibers serve as water transport channels. The CCP works concurrently as a volume light absorber and a secondary liquid transport system (Figure 10.6a). The CCP behaves as a strong light absorber in the visible range from 400 nm to 800 nm, achieving absorption values up to 98.5%. Morphology analysis via SEM demonstrates micro-sized porous channels ranging from hundreds of nanometers to tens of microns (Figure 10.6b, inset). When applied in liquid absorption measurements, the CCP with cotton fibers device shows an enhanced filling factor, i.e., the

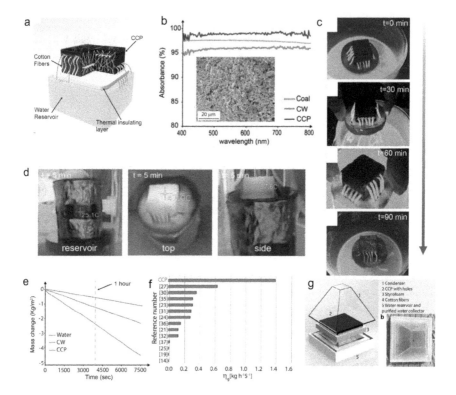

FIGURE 10.6 Clean carbon cycle with natural coal for low-cost solar water evaporation. (a) Schematic of the solar absorber structure consisting of carbonized compressed powder (CCP) and cotton fibers. (b) Absorption spectra of fossil charcoal, CW (carbonized wood) versus CCP (inset: cross-sectional SEM images of CCP). (c) Dynamic liquid absorption in the device composed of CCP and the complex network of cotton fibers over 90 min. The system reaches saturation in half the time than CCP with no fibers. (d) Spatial thermal images and temperature distributions of the desalination device with CCP. (e) Mass loss rate of water versus time under one sun illumination. (f) Comparison of global efficiency ηg defined as kg of fresh water produced per every hour, per every dollar ($) at one sun intensity of solar desalination devices from literature. (g) Large-scale system prototype with CCP-engineered solar evaporation devices. (Reproduced from Mazzone et al. [9], with permission.)

extent that the device is filled with water, of 100% in half of the saturation time compared with CCP with no fibers (Figure 10.6c). Such a CCP-engineered system provides heat localization properties (Figure 10.6d) with 47 °C on the top surface and 37 °C on the side when illuminated with one sun. The temperature of the evaporator surface is 88% higher than that of bulk water when illuminated with the same source. In solar evaporation experiments, the 3D CCP-engineered evaporator reports a water evaporation rate of up to 2.2 kg m^{-2} h^{-1} under 1 sun, exceeding the theoretical limit of 1.47 kg m^{-2} h^{-1} of 2D evaporators [45, 46].

The authors of [9] provide a comparative analysis of various solar desalination technologies, including materials cost. They report the CCP-engineered system with the highest performance in terms of evaporation rate per unit area, unit time, and unit material cost of 1.39 kg h^{-1}\$$^{-1}$ under one sun (Figure 10.6f). The authors perform outdoor experiments with the implementation of a large-scale CCP prototype (34 × 34 cm) in seawater, reporting the generation of drinkable water with salinity and ion concentrations within the requirements of WHO. This work estimates that an 8 m^2 area CCP evaporation system can serve a family of four people, generating freshwater at the cost of around 0.002 \$ L^{-1}, comparable to the average cost from reverse osmosis technique (0.0015 \$ L^{-1}).

Because solar vapor generation (SVG) is an energy-intensive process requiring an evaporation enthalpy of 2450 J g^{-1}, the theoretical evaporation rate for a regular 2D evaporator is limited to 1.6 kg m^{-2} h^{-1} under natural sunlight [30]. Reducing the energy requirements of SVG is another pathway to enhance the solar water purification performances of different materials. Zhou and co-workers introduced hydratable polymer networks to tune the water states and partially activate the water, thus lowering the energy barrier and increasing water evaporation [30]. The authors apply a light-absorbing hydrogel with hydratable polymer networks (h-LAH), fabricated by infiltrating polypyrrole (PPy) absorber into polyvinyl alcohol (PVA) and chitosan matrix (Figure 10.7a). Due to the hydration effect, the hydrogel polymer binds with nearby water molecules through hydrogen bonding, reducing the vaporization energy compared to bulk water (Figure 10.7a). The prepared h-LAHs are black and flexible (Figure 10.7b), with interconnecting porous structures throughout the hydrogel skeleton (Figure 10.7c). Three types of water exist in the hydrogel networks: bound water (Figure 10.7d, region I), intermediate water (Figure 10.7d, region II), and free water (Figure 10.7d, region III). Each state possesses distinct phase change behavior with different energy transfer

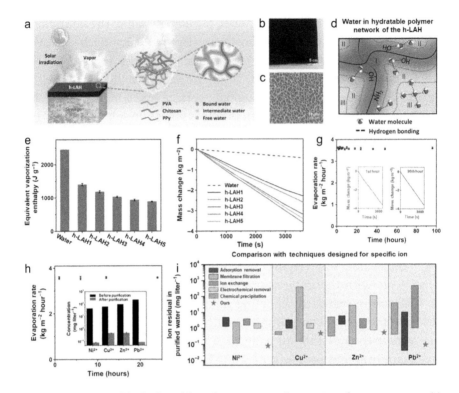

FIGURE 10.7 Highly hydratable polymer networks tuning the water state. (a) Schematic illustration of solar vapor generation (SVG) based on the h-LAH. (b) Photograph of as-prepared h-LAH sample. (c) SEM image of the micron-sized pores in the freeze-dried h-LAH. (d) Schematic of the tuned water bonding states in the hydratable polymer network of the h-LAH. (regions I-III marked with dash lines represent the bound, intermediate, and free water, respectively.) (e) The equivalent water vaporization enthalpy of bulk water and water in h-LAH polymer. (f) The mass loss of pure water for different h-LAH samples under one sun compared with pure water. (g) The duration test of the h-LAH4 on continuous solar desalination for 96 hours. Insets show mass water loss with the h-LAH4 at the 1st hour and 96th hour. (h) Purification of heavy metal polluted water. Inset: concentration of heavy metal ions in solution before/after purification. (i) Ion residual in purified water compared with several competitive purification techniques designed for a specific ion. (Reproduced from Zhou *et al.* [30], with permission.)

pathways. The authors measure the equivalent water vaporization enthalpy (E_{equ}) of other h-LAH hydrogels and bulk water as $E_{equ} = E_0 \, m_0/m_g$, with E_0 and m_0 the vaporization enthalpy and mass change of bulk water, respectively, and m_g the mass change in h-LAHs. The reduced E_{equ} in h-LAHs compared with bulk water confirms that intermediate water decreases the energy barrier for water vaporization (from 2450 J g^{-1} to 881 J g^{-1}). Hydrogel networks effectively tune the water states (Figure 10.7e). The h-LAH4 shows optimized water vaporization enthalpy and water content in polymer networks, exhibiting an evaporation rate of 3.6 kg m^{-2} h^{-1} and energy efficiency up to 92% under one sun (Figure 10.7f). This hydrogel-based evaporator also shows stability under continuous 1-sun illumination over 96 hours while maintaining a stable evaporation rate in the operating time (Figure 10.7g). When applied in simulated industrial wastewater containing mixed heavy metal ions, including Ni^{2+}, Cu^{2+}, Zn^{2+}, and Pb^{2+}, the h-LAH evaporator reduces the metal ion concentrations to below 1 mg L^{-1} after purification (Figure 10.7h). This purification performance of h-LAH in wastewater with heavy metal ions is comparable to commercialized techniques designed for specific ions (Figure 10.7i).

10.2.2 Metamaterials with Interface Engineering for Enhancing the Evaporation Rate

Solar water desalination involves multiple mass/energy transport processes at the interfaces [8, 16]. These processes include water transport between liquid/solid materials interface, heat transport in the light absorber to nearby air/liquid interfaces, and vapor generation at the material/air interface. Interface engineering for accelerating mass transfer and energy utilization is another efficient pathway to enhance the water evaporation rate [47–51].

Zhou *et al.* report a hybrid hydrogel evaporator consisting of a hydrophilic polymer framework (polyvinyl alcohol, PVA) and a solar absorber (reduced graphene oxide, rGO) that achieves an evaporation rate of 2.5 kg m^{-2} h^{-1} under one sun illumination [26]. The polymer networks reduce the water evaporation enthalpy, and the rGO enables efficient light absorption with abundant capillary channels for water transport (Figure 10.8a). In contrast, the interpenetrated rGOs in the polymer networks confine the heat to the nearby molecular meshes. The hybrid hydrogel evaporator possesses porous structures with capillary channels throughout the polymer skeleton (Figure 10.8b). The UV-Vis-NIR absorption spectrum shows nearly 100% light absorption over the full solar spectrum from 250 nm to 2500 nm (Figure 10.8c). The authors measure temperature variation and report rapid heating in the obtained hydrogel with capillarity facilitated water transport, denoted as CTH, with an equilibrated temperature of 32 °C in 10 min (Figure 10.8d). Leveraging such optimized water transport channels, the CTH shows the evaporation rate of 2.5 kg m^{-2} h^{-1} with energy efficiency up to 95% under one sun illumination (Figure 10.8e). Infrared

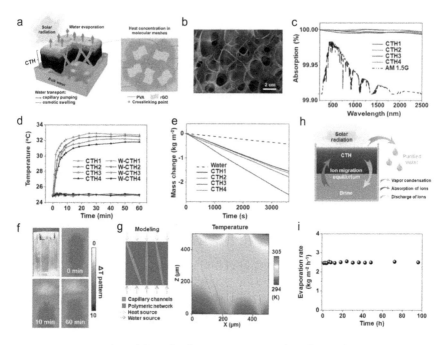

FIGURE 10.8   Hydrogel-based solar evaporator with enhanced water transport and antifouling properties. (a) Schematic illustration of solar vapor generation based on hybrid hydrogels with capillarity facilitated water transport (CTH). (b) SEM image of micron-sized pores of the CTH. (c) UV-Vis-NIR spectra of CTH sheets (Thickness: ~1 mm). Black dashed line: normalized spectral solar irradiance density of AM 1.5 G solar spectrum. (d) Time evolution of the temperature of the evaporation surface in the CTHs and bulk water (W-CTH1, 2, 3, and 4) under one sun. (e) Mass water loss for different CTH samples under one sun compared with pure water. (f) Setup for the solar vapor generation and infrared images of the CTH sample at different times. (g) Theoretical simulation of the temperature distribution near the evaporation surface of CTH, where the bright ribbon-like areas and dark background represent the capillary channels and polymeric network, respectively. (h) Schematic illustration of solar desalination based on CTH. (i) Duration test of the CTH for continuous solar desalination over 96 h under one sun. (Reproduced from Zhou *et al.* [26], with permission.)

images visualize the temperature changes in the evaporation setup based on CTH, illustrating localized heating in the top CTH evaporators (Figure 10.8f). Structural modeling simulates heat transfer in the capillary channels and polymer networks, predicting a maximum temperature in the CTH of 305 K (~32 °C) (Figure 10.8g). The authors report experiments with the CTH-based evaporator for seawater desalination (Figure 10.8h), decreasing four orders of magnitude in salinity for the purified water.

CTH desalination devices show stable operation at initial evaporation rates over 96 hours in seawater with salinity reaching 20% (Figure 10.8i).

Besides tuning the water bonding states in the materials, controlling water contents at the material/vapor interface is a possible route to enhance the solar evaporation rate. Li *et al.* explored this avenue by developing a covalent organic framework (COF)/graphene dual-region hydrogel (CGH) for solar water evaporation [20]. This evaporator contains both hydrophobic and hydrophilic regions. The COF-loaded reduced graphene oxide (COF@rGO) contributes to the hydrophilic surface. At the same time, the pure rGO region possesses hydrophobic properties (Figure 10.9a). The authors synthesize COF-SO$_3$H powders by introducing sulfonic acid groups to COF through a hydrothermal method (Figure 10.9b). The CGH samples have extensive microporous structures, which work as capillary channels for water transport (Figure 10.9c). The water contact angle decreases from 130.6 ± 4.6° to 69.7 ± 1.7° with the rising coverage of COF-SO$_3$H in rGO, showing enhanced hydrophilicity by introducing COF-SO$_3$H (Figure 10.9d). Moreover, CGH-based evaporators present a higher dark evaporation rate of up to 0.50 kg m^{-2} h^{-1} than pure water (~0.13 kg m^{-2} h^{-1}), with evaporation enthalpy decreasing from 2450 J g^{-1} in bulk water to 1043 J g^{-1} in CGHs (Figure 10.9e). Evaporation rate and enthalpy changes show a reduced energy barrier for water vaporization with hydrophilic/hydrophobic dual-region hydrogels [20]. The authors construct the water evaporation system by attaching the CGH on a polystyrene (PS) heat-insulating foam (Figure 10.9f). The CGH-based systems show faster water evaporation rates than bulk water (Figure 10.9g), reporting an evaporation rate up to 3.69 kg m^{-2} h^{-1} and desalination efficiency of ~92% (Figure 10.9h). CGH evaporator maintains stable evaporation performance in simulated seawater (3.45 kg m^{-2} h^{-1}) over 7 cycles of 10 h each (Figure 10.9i). The authors extend this CGH-based evaporator to purify the sewage with methyl orange and report a removal efficiency of 100%, proving its application for dye wastewater purification (Figure 10.9j).

Salt accumulation at the evaporation surface limits the stable operation of water desalination devices, especially in high-salinity brine (>10%) [25]. To address this problem, He *et al.* engineer a nature-inspired bimodal porous structure (e.g., balsa wood) as a salt-accumulation-free solar evaporator for high-salinity brine desalination [25]. In the bimodal porous structures, the microchannels provide capillary transport for water (Figure 10.10a). The surrounding large vessel channels allow diffusion and convection of high-concentration brine, avoiding the formation of salt

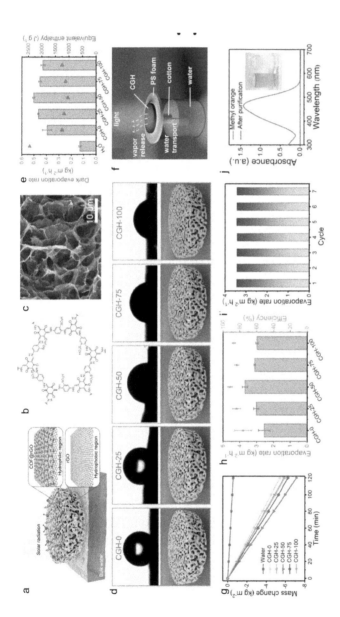

FIGURE 10.9 Hydrophilic and hydrophobic dual-region hydrogel for water vaporization. (a) Scheme of the COF/graphene dual-region hydrogel for solar water evaporation. (b) Chemical structure of COF-SO₃H. (c) SEM images of CGH-50. (d) Water contact angles of the freeze-dried samples and corresponding schematic drawings of the dual regions in the CGHs. (e) Dark evaporation rate and evaporation enthalpy of pure water and water in CGHs. (f) Scheme of the setup for the water evaporation test. (g) Time-dependent mass change of pure water and CGHs under one sun irradiation (1 kW m⁻²). (h) Water evaporation rate from CGHs and energy efficiency under one sun. (i) Stability test applying simulated seawater and CGH-50 over seven cycles. (j) UV-vis spectra and photographs of dye-contaminated water and purified water. (Reproduced from Li *et al.* [20], with permission.)

crystals on the surface of the evaporator. The authors carbonized the surface of natural balsa wood to fabricate the bimodal evaporator (Figure 10.10b). The 3D interconnected porous structures of natural balsa wood remain after carbonization, ensuring sufficient brine supply during desalination (Figure 10.10c). The carbonized bimodal surface (around 2.5 mm in thickness) shows broadband light absorption of nearly 97% across the full solar spectrum, serving as an efficient solar absorber for photothermal conversion (Figure 10.10d). When applied in simulated high salinity brine (15 wt%), the evaporation rate of this bimodal evaporator increases from 0.8 kg m^{-2} h^{-1} under one sun to 6.4 kg m^{-2} h^{-1} under six sun (Figure 10.10e). Correspondingly, the evaporation efficiency rises from ~57% to ~78% (Figure 10.10f). The bimodal evaporator maintains its evaporation rate in accelerated durability tests under six suns. However, the evaporation rate of the control group composed of PDMS/balsa and unimodal cedar decreases by 29% and 18%, respectively, due to the severe salt accumulation (Figure 10.10g). The bimodal evaporator presents long-term cycling performance for 20 days (7 h per day) with no appreciable salt accumulation on the surface (Figure 10.10h). The authors examine the salt rejection ability of this bimodal evaporator by directly adding 1 gram of solid NaCl on the top surface. They report the salts to dissolve entirely back to the brine underneath after 7 h illumination with two suns intensity (Figure 10.10i).

Tuning wetting states at the evaporator/vapor interface is another strategy to accelerate the water evaporation rate. Guo *et al.* report a surface-wetting states chemical tuning on hydrogel-based evaporators to enhance the water evaporation performances [19]. The authors introduce hydrophobic patches to the hydrophilic surface of hydrogel via partial modification with trichloro(octadecyl)silane (OTS). This process generates island-shaped patches on the hydrophilic evaporator (Figure 10.11a). Such manufactured surface-wetting states increase the water thickness in the hydrophilic region, with the hydrogel networks providing rapid water transport to the evaporation surface. Contact angle increases from 0° at 0% OTS to 130.7° at 90% OTS, demonstrating enhanced hydrophobicity after OTS surface modification on hydrogels (Figure 10.11b). Infrared images illustrate the confined heat energy at the top surface of the hydrogel evaporator during 60 min illumination under 1 sun (Figure 10.11c). The patchy-surface hydrogel (PSH) introduced in this work shows an enhanced vaporization rate of up to ~4.0 kg m^{-2} h^{-1}, with an energy efficiency of ~93% under one sun (Figure 10.11d-e). Molecular dynamics simulations report that

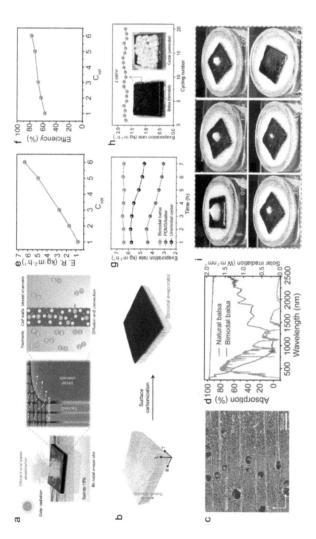

FIGURE 10.10 Bimodal porous solar evaporator engineered from balsa wood. (a) Scheme of the microstructures and working principle of the bimodal porous balsa wood for high salinity brine desalination. (b) Digital images of natural balsa and the salt-resistant bimodal evaporator. (c) Top-view SEM image for numerous wide vessel channels and narrow tracheids along the tree growth direction. (d) Absorption spectra of natural balsa and the bimodal wood. (e) The bimodal porous wood evaporator's evaporation rates and (f) evaporation efficiencies under various illumination intensities of 1–6 suns. (g) Evolution of the evaporation rate for the bimodal and unimodal evaporators (PDMS/ balsa and cedar evaporators). (h) Endurance measurement results of the bimodal evaporator exposed to 2 suns for 20 days, with seven h per day; insets: comparison of salt accumulation on the surface of the bimodal evaporator and unimodal evaporator). (i) Salt rejection capability from the surface of the bimodal evaporator under two suns illumination. (Reproduced from He et al. [25], with permission.)

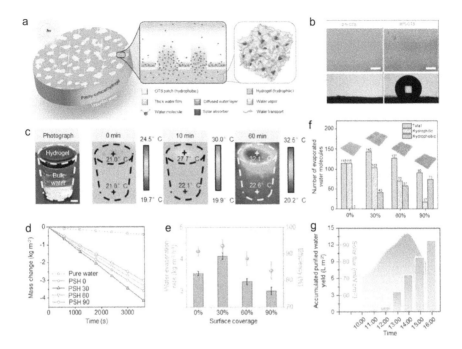

FIGURE 10.11 Surface wetting states tuning. (a) Schematic of patchy-surface hydrogels (PSHs) for enhanced solar-driven evaporation. (b) Optical images and evaporation state contact angles of PSHs with 0% and 90% OTS-covered surfaces. (c) Photographs of the setup for the solar-driven vapor generation test using PSH 30 and the infrared images of the sample over time. Scale bar: 1 cm. (d) Mass change of water and (e) evaporation rate and corresponding energy efficiency of the PSH under one sun. (f) Evaporated water molecules with contributions from the hydrophilic and hydrophobic regions for different OTS surface coverages after the same evaporation duration. (g) Purified water yield from seawater using PSH 30 under natural sun irradiation. (Reproduced from Guo *et al.* [19], with permission.)

PSH with 30% OTS coverage provides the largest molecular water evaporation number under the same energy input (Figure 10.11f). Simulation outcomes agree with experimental results in Figure 10.11d–e, predicting how the engineered surface wetting states enhance the vaporization behavior of water. Outdoor experiments, conducted from 10 am to 4 pm with averaged sunlight intensity of ~0.75 kW m^{-2}, report an average water purification rate of 2.1 kg m^{-2} h^{-1}. The system reduces the seawater content of primary metal ions (Na$^+$, Mg$^+$, K$^+$, and Ca$^+$) by ~3 orders of magnitude (Figure 10.11g).

10.2.3 Metamaterials Structural Engineering

Evaporator structural and shape engineering is also a practical pathway to enhance water evaporation efficiency while preventing salt accumulation on the evaporator [17, 21, 23, 52, 53]. Recently, Yao *et al.* report the design of a Janus-interface solar-steam generator (J-SSG), which decouples the photothermal conversion from water vaporization on different sides [54]. Unlike classical interfacial evaporators that use a single surface for solar-thermal conversion and water evaporation simultaneously, the Janus-interface evaporator exploits two different sides with asymmetric structures. One side works as the solar absorber, and the other for water vaporization. A Janus interface eliminates the destructive interference between incident light and water vapor escaping from the evaporation surface, which occurs in classical evaporators (Figure 10.12a), and lowers the energy barrier for water evaporation. The authors manufacture the light absorber side by fast-laser processing on Janus copper foil, coating the other side with acetylated chitosan and PVA aerogel (ACPA) mixture acting as a porous evaporation surface (Figure 10.12b) (Figure 10.12b). Pyramid-like arrays with ~50 μm in period obtained after fast-laser treatment enable multiple reflections of sunlight within the structures and thereby enhance the light absorption by trapping the light (Figure 10.12c, d), reporting ~96% of the incident light from 250 nm to 2500 nm (Figure 10.12cm d). The polymer networks tune the bonding states of surrounding water molecules and reduce the water vaporization enthalpy from 2346 J g^{-1} in bulk water to 1532 J g^{-1} in ACPA (Figure 10.12e, f). Experiments with a J-SSG system under one sun irradiation report an evaporation rate of 2.21 kg m^{-2} h^{-1}, for an evaporator with a large area of 100 cm^2 (Figure 10.12g).

The spatial separation between light absorption and vapor generation enables opportunities to optimize each side. In the work [54], the authors report almost full use of solar energy in evaporation, obtaining a water evaporation rate of 3.88 kg m^{-2} h^{-1} under two sun illumination (Figure 10.12h).

Finite element simulations predict a large humidity gradient up to 50% between the J-SSG evaporator and metal condenser (Figure 10.12i). This large humidity gradient facilitates the water vapor condensation into liquid water, contributing to the observed solar evaporation efficiency of 88%.

Avoiding salt accumulation on the evaporation surface is crucial to maintain light absorption and evaporation rates [23, 53, 55]. Recently, Wang *et al.* engineered a metal–phenolic network (MPN) into a 3D

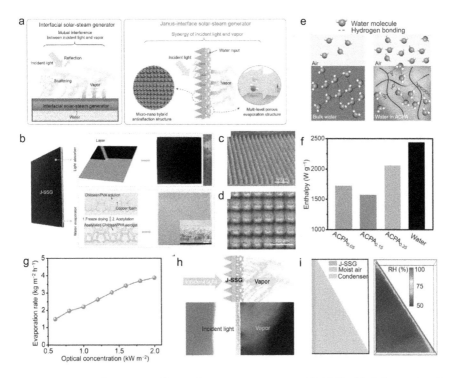

FIGURE 10.12   Janus-interface solar-steam generator (J-SSG). (a) Schematic of the steam generation process of the solar-steam generator with the traditional interface and Janus-interface. (b) The schematic fabrication process of the solar absorption and evaporation surfaces for J-SSG. (c-d) SEM images of the solar absorption surface. (e) Schematic of the water molecules in bulk water and hydratable polymer network of ACPA. (f) Calculated evaporation enthalpy of pure water and water in ACPA at different proportions of PVA/chitosan. (g) Evaporation rate of J-SSG under increasing solar irradiation. (h) Schematic diagram of fast solar-thermal water evaporation in J-SSG and the photograph of the J-SSG generating steam. (i) Theoretical simulation of the solar desalination system (Left side: condensation system; right side: balanced humidity distribution between evaporator and condenser). (Reproduced from Yao *et al.* [54], with permission.)

evaporator with directional salt crystallization and no liquid discharge [56]. The implemented 3D evaporator consists of photothermal super hydrophilic/superhydrophobic sponges and hydrophilic threads (Figure 10.13a). The MPN modification on the sponge enables photothermal conversion and channels for evaporation. At the same time, the side-twining threads allow site-selective crystallization of salts. The authors use contact angle measurements to prove the super hydrophilic nature of the original

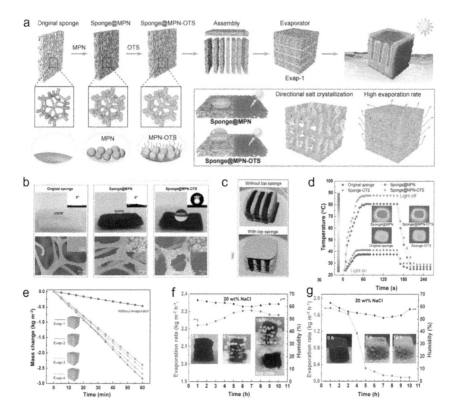

FIGURE 10.13   3D evaporator engineering for directional salt crystallization. (a) Schematic of the modification of sponges by MPNs and their assembly into 3D evaporators for solar desalination (Right corner: scheme of photothermal and selective wettability properties of MPN-engineered sponges.) (b) Photographs and SEM images of original sponge, sponge@MPN, and sponge@MPN-OTS. (c) Photographs of the immersion of Evap-1 with or without a top sponge cover in fluorescent aqueous solution under UV light. (d) Thermal infrared images of different sponges in the presence or absence of solar irradiation in air. (e) Evaporated water vs. time. Evaporation rates of water during treatment of 20 wt% brine using Evap-1 (f) or Evap-5 (g). Insets: photographs of evaporators with salt crystals at different treatment times. (Reproduced from Wang *et al.* [56], with permission.)

melamine sponge and super hydrophobic property after modification with the octadecyltrimethoxysilane (OTS) (Figure 10.13b). The sponge skeleton's porous structure provides many capillary channels for water transport during solar evaporation (Figure 10.13b). The fluorescent images of the aqueous solution visualize the water flow and distribution in the evaporator with/without the top sponge (Figure 10.13c), illustrating the selective water transport in this hybrid hydrophilic/hydrophobic evaporator (Figure 10.13c). The surface temperatures of sponge@MPN (~80 °C) and sponge@MPN-OTS (~90 °C) are higher than original sponge (~37 °C), demonstrating strong photothermal light-matter interactions in MPN (Figure 10.13d). In these results, the 10 °C temperature increase of sponge@MPN-OTS compared to sponge@MPN proves the surface hydrophobicity of the metamaterial sponge enhances its photothermal effects (Figure 10.13d). The evaporation rate of the sponge-based evaporator (Evap-1) reaches 2.84 kg m^{-2} h^{-1} under 1 sun, showing a 500% increment to bulk water (0.47 kg m^{-2} h^{-1}) (Figure 10.13e). The studied Evap-1 also shows a stable evaporation rate of 2.2–2.3 kg m^{-2} h^{-1} over 10 hours with selective salt crystallization on the side threads (Figure 10.13f). The authors report that, in comparison, the evaporation rate of pure super hydrophilic sponge decays in 3 hours, with the top surface fully covered by salt crystals (Figure 10.13g).

Along a similar research direction, Wu *et al.* developed a self-rotating solar evaporator with separated high-temperature and low-temperature zones to address salt accumulation, reporting an evaporation rate of 2.6 kg m^{-2} h^{-1} under one sun [55]. Device manufacturing starts from PS spheres and follows coating with α-cellulose fibers to increase the surface roughness. The process then includes the use of a polydopamine (PDA) layer to enhance the hydrophilicity and a polypyrrole (PPy) layer to improve the light absorption (Figure 10.14a). Thanks to a spherical shape and lower density than water, this 3D evaporator rotates spontaneously in response to the weight imbalance and refreshes the evaporation surface (Figure 10.10a-c).

The authors introduce PDA to the PS sphere and report a superhydrophilic surface with a water contact angle decreasing from above 90° to 0°. (Figure 10.10c, inset). The top surface temperature of this spherical evaporator (dry state) reaches up to around 50 °C within 1 min under one sun illumination (Figure 10.10d). Infrared images show heat confinement within the top part of the evaporator (Figure 10.10d, inset), reporting a large temperature difference of ~18 °C between the top and bottom regions of the evaporator. Temperature distribution images present high-temperature and low-temperature regions with a temperature difference of ~15 °C on the spherical evaporator under illumination, with the low-temperature region accounting for 37.2–41.8% of the total surface (Figure 10.10e). The weight imbalance due to salt accumulation triggers the self-rotation of the spherical evaporator. It cleans its top surface for evaporation (Figure 10.10f).

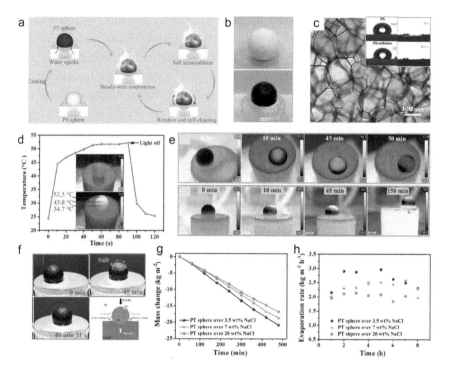

FIGURE 10.14 Self-rotating photothermal evaporator with dual temperature evaporation zones. (a) Fabrication process of the photothermal sphere and the working mechanism of salt-resisting solar evaporation. (b) Digital photographs of the PS-cellulose sphere (upper) and PS-photothermal sphere floating on water (bottom). (c) Optical image of the macro-porous inner cavity structure of the PS sphere. Insets: Static contact angle of the water droplet on PS and PS-cellulose surfaces (left); time-lapse snapshots of the water droplet absorption by the PS-cellulose-PPy-PDA. (d) Time-dependent temperature evolution of the PS-photothermal sphere (dry state) under one sun irradiation. Insets: IR image of the PS-photothermal sphere without light irradiation (upper) and after irradiation for 1.5 min (down). (e) IR images of the photothermal sphere over 20 wt% NaCl solution upon one sun irradiation over time. (f) Time-lapse snapshots of the salt accumulation on the photothermal sphere, induced rotation, and rotation mechanism. (g) Time-dependent mass loss of water under one sun irradiation. (h) Evaporation rate of the photothermal sphere over water sources with different Na^+ concentrations. (Reproduced from Wu et al. [55], with permission.)

The evaporation rate declines slightly from 2.6 kg m^{-2} h^{-1} to 2.06 kg m^{-2} h^{-1} in 8 h continuing tests under 1 sun, with salt concentration increasing from 3.5 wt% to 20 wt% (Figure 10.10g, h). The evaporation rates reported with different salt concentrations are four times higher than that of a bulk salt solution under the same illumination.

10.3 SUMMARY AND OUTLOOK

In this chapter, we summarized recent research advances driven by the applications of metamaterials in solar desalination. The research focused on the key areas of materials, interface, and structural engineering to improve solar energy conversion efficiency and the operational stability of solar evaporation systems.

The work of Zhou *et al.* [24] optimized nanostructuring in Al-based plasmonic evaporators, reporting broadband solar absorption over 96% and water evaporation rate of ~1.0 kg m^{-2} h^{-1} under one sun. Li *et al.* [20] demonstrated the importance of tuning the water contents and evaporation enthalpy by designing hydrophilic/hydrophobic hybrid surface regions, reporting over 57% decreased water evaporation enthalpy and evaporation rate of 3.69 kg m^{-2} h^{-1}. Yao *et al.* [54] highlighted the importance of structural engineering by spatially decoupling light absorption and vapor generation through a Janus-interface, reporting water evaporation rates as high as 3.88 kg m^{-2} h^{-1} under two sun and a solar evaporation efficiency of 88%. The work of He *et al.* [25] provided an example of avoiding salt accumulation during solar desalination by designing a nature-inspired bimodal porous structure.

Despite many achievements, further opportunities exist to improve this technology for future challenges. The first aspect is enhancing the water collection rate. Even though the water evaporation rate in current research reached over 10 kg m^{-2} h^{-1} under one sun [21], the water collection efficiency in many cases is limited to ~35%, which is equivalent to a water collection rate of ~0.5 kg m^{-2} h^{-1} [17, 57]. The main challenges in improving this performance are high optical losses, up to 35%, from the condensed vapor in the system [17], and low thermal conductivity of the top transparent cover made of polymer or glass [58]. The inverted evaporator design in the work of Wang *et al.* [17] proposed a solution to mitigate optical losses, reporting a water collection rate of 1.063 kg m^{-2} h^{-1} and overall efficiency of solar-to-collected water of ~70%. We expect future research to further enhance this result by proposing novel structural engineering of the evaporator. Harnessing latent energy from the environment is another direction to improve current desalination technologies further. These include, e.g., collecting radiational heat energy from the environment and the convective air flow from the wind to enhance evaporation performances [59].

Coupling with other systems and optimizing energy management is another aspect of future research to improve solar conversion efficiency.

Since solar energy is only available in the daytime with variable sunlight intensity, combining solar evaporators with reversible phase change materials could be a promising way to store the excess thermal energy in the day and release it at night for evaporation [60].

Another important direction is broadening the applications of solar desalination systems to other areas, including, e.g., wastewater treatments. In industrial and domestic wastewater, organic dyes or bacteria remain appreciable in the bulk water reservoir after purification. Dyes and bacteria can block water transportation in the evaporator and deteriorate the long-term stability of the evaporation system. Coupling solar evaporation and photocatalytic hydrogen peroxide (H_2O_2) or hydroxyl radical ($HO^{\bullet}$) generation systems can actively degrade the dyes and kill the bacteria, significantly improving the performances of solar evaporation systems for industrial wastewater treatment.

Materials cost and scalability are two key factors for large-scale implementation of solar desalination. The work of Mazzone et al. [9] provided a comparative analysis of the desalination rate per material cost for various desalination technologies. They reported that a natural coal and cotton-engineered evaporator showed the highest performance of 1.39 kg h^{-1} \$$^{-1}$ under one sun, generating freshwater at the cost of around 0.002 \$ L^{-1}, which is comparable with reverse osmosis. We expect future works to focus on exploring additional low-cost materials and structures.

Long-term stability against different environmental conditions is another crucial factor for practical implementations of solar desalination. The current research focused on addressing the salt accumulation problem on the evaporator by using the self-rotating metamaterial design [55], bimodal porous structures [25], and directional salt crystallization [56]. Future work could explore additional environmental issues, such as, e.g., the instability of organic polymers to UV light, increasing the stability of current systems [19].

10.4 APPENDIX

10.4.1 Solar Steam Efficiency

The solar steam efficiency η can be calculated by the following formula [22, 34, 61]:

$$\eta = \dot{m} h_{LV} / C_{opt} P_0 \qquad (10.1)$$

where $\dot{m}(t) \equiv dm/dt$ is the mass flux; h_{LV} is the latent enthalpy of the liquid-vapor phase change; P_0 is the solar irradiation power of 1 sun (1 kW m^{-2}) and $C_{opt}P_0$ represents the illumination intensity on the absorber surface.

10.4.2 Desalination Rate Per Material Cost

The device performance of solar evaporators, including materials cost, is essential for industrial applications. The global efficiency taking into account the materials cost, can be calculated with the following formula [9]:

$$\eta_g = \frac{\text{kilograms of H}_2\text{O evaporated}}{\text{hour} \times \$ \times \text{sun illumination intensity}} \quad (10.2)$$

REFERENCES

[1] Xinyu Chen, Yaxing Liu, Qin Wang, Jiajun Lv, Jinyu Wen, Xia Chen, Chongqing Kang, Shijie Cheng, and Michael B McElroy. Pathway toward carbon-neutral electrical systems in china by mid-century with negative co2 abatement costs informed by high-resolution modeling. *Joule*, 5(10):2715–2741, 2021.

[2] Srinivas Garimella, Kristian Lockyear, David Pharis, Omar El Chawa, Matthew T Hughes, and Girish Kini. Realistic pathways to decarbonization of building energy systems. *Joule*, 6(5):956–971, 2022.

[3] Jan-Georg Rosenboom, Robert Langer, and Giovanni Traverso. Bioplastics for a circular economy. *Nature Reviews Materials*, 7(2):117–137, 2022.

[4] Seda Sarp, Santiago Gonzalez Hernandez, Chi Chen, and Stafford W Sheehan. Alcohol production from carbon dioxide: methanol as a fuel and chemical feedstock. *Joule*, 5(1):59–76, 2021.

[5] David W Keith, Geoffrey Holmes, David St Angelo, and Kenton Heidel. A process for capturing co2 from the atmosphere. *Joule*, 2(8):1573–1594, 2018.

[6] Ilkka Hannula and David M Reiner. Near-term potential of biofuels, electrofuels, and battery electric vehicles in decarbonizing road transport. *Joule*, 3(10):2390–2402, 2019.

[7] Tingting Zheng, Menglu Zhang, Lianghuan Wu, Shuyuan Guo, Xiangjian Liu, Jiankang Zhao, Weiqing Xue, Jiawei Li, Chunxiao Liu, Xu Li, et al. Upcycling co2 into energy-rich long-chain compounds via electrochemical and metabolic engineering. *Nature Catalysis*, 5(5):388–396, 2022.

[8] Chaoji Chen, Yudi Kuang, and Liangbing Hu. Challenges and opportunities for solar evaporation. *Joule*, 3(3):683–718, 2019.

[9] Valerio Mazzone, Marcella Bonifazi, Christof M Aegerter, Aluizio M Cruz, and Andrea Fratalocchi. Clean carbon cycle via high-performing and low-cost solar-driven production of freshwater. *Advanced Sustainable Systems*, 5(10):202100217, 2021.

[10] UN water. (2018). Water scarcity. http://www.unwater.org/water-facts/scarcity/

[11] Water stress by country in 2040. https://reliefweb.int/map/world/water-stress-country-2040

[12] Global horizontal solar irradiation map. https://solargis.com/maps-and-gis-data/download/world

[13] Nathan S Lewis. Research opportunities to advance solar energy utilization. *Science*, 351(6271):aad1920, 2016.

[14] Taikan Oki and Shinjiro Kanae. Global hydrological cycles and world water resources. *Science*, 313(5790):1068–1072, 2006.

[15] Peng Tao, George Ni, Chengyi Song, Wen Shang, Jianbo Wu, Jia Zhu, Gang Chen, and Tao Deng. Solar-driven interfacial evaporation. *Nature Energy*, 3(12):1031–1041, 2018.

[16] Fei Zhao, Youhong Guo, Xingyi Zhou, Wen Shi, and Guihua Yu. Materials for solar-powered water evaporation. *Nature Reviews Materials*, 5(5):388–401, 2020.

[17] Fengyue Wang, Ning Xu, Wei Zhao, Lin Zhou, Pengcheng Zhu, Xueyang Wang, Bin Zhu, and Jia Zhu. A high-performing single-stage invert-structured solar water purifier through enhanced absorption and condensation. *Joule*, 5(6):1602–1612, 2021.

[18] Ying Zhang, Yan Wang, Bin Yu, Kuibo Yin, and Zhonghua Zhang. Hierarchically structured black gold film with ultrahigh porosity for solar steam generation. *Advanced Materials*, 34(21):2200108, 2022.

[19] Youhong Guo, Xiao Zhao, Fei Zhao, Zihao Jiao, Xingyi Zhou, and Guihua Yu. Tailoring surface wetting states for ultrafast solar-driven water evaporation. *Energy & Environmental Science*, 13(7):2087–2095, 2020.

[20] Changxia Li, Sijia Cao, Jana Lutzki, Jin Yang, Thomas Konegger, Freddy Kleitz, and Arne Thomas. A covalent organic framework/graphene dual-region hydrogel for enhanced solar-driven water generation. *Journal of the American Chemical Society*, 144(7):3083–3090, 2022.

[21] Jinlei Li, Xueyang Wang, Zhenhui Lin, Ning Xu, Xiuqiang Li, Jie Liang, Wei Zhao, Renxing Lin, Bin Zhu, Guoliang Liu, et al. Over 10 kg m- 2 h- 1 evaporation rate enabled by a 3d interconnected porous carbon foam. *Joule*, 4(4):928–937, 2020.

[22] Changxu Liu, Jianfeng Huang, Chia-En Hsiung, Yi Tian, Jianjian Wang, Yu Han, and Andrea Fratalocchi. High-performance large-scale solar steam generation with nanolayers of reusable biomimetic nanoparticles. *Advanced Sustainable Systems*, 1(1-2):1600013, 2017.

[23] Lei Wu, Zhichao Dong, Zheren Cai, Turga Ganapathy, Niocholas X Fang, Chuxin Li, Cunlong Yu, Yu Zhang, and Yanlin Song. Highly efficient three-dimensional solar evaporator for high salinity desalination by localized crystallization. *Nature Communications*, 11(1):521, 2020.

[24] Lin Zhou, Yingling Tan, Jingyang Wang, Weichao Xu, Ye Yuan, Wenshan Cai, Shining Zhu, and Jia Zhu. 3d self-assembly of aluminium nanoparticles for plasmon-enhanced solar desalination. *Nature Photonics*, 10(6):393–398, 2016.

[25] Shuaiming He, Chaoji Chen, Yudi Kuang, Ruiyu Mi, Yang Liu, Yong Pei, Weiqing Kong, Wentao Gan, Hua Xie, Emily Hitz, et al. Nature-inspired salt resistant bimodal porous solar evaporator for efficient and stable water desalination. *Energy & Environmental Science*, 12(5):1558–1567, 2019.

[26] Xingyi Zhou, Fei Zhao, Youhong Guo, Yi Zhang, and Guihua Yu. A hydrogel-based antifouling solar evaporator for highly efficient water desalination. *Energy & Environmental Science*, 11(8):1985–1992, 2018.

[27] Lin Zhou, Yingling Tan, Dengxin Ji, Bin Zhu, Pei Zhang, Jun Xu, Qiaoqiang Gan, Zongfu Yu, and Jia Zhu. Self-assembly of highly efficient, broadband plasmonic absorbers for solar steam generation. *Science advances*, 2(4):e1501227, 2016.

[28] Yusuf Shi, Renyuan Li, Yong Jin, Sifei Zhuo, Le Shi, Jian Chang, Seunghyun Hong, Kim-Choon Ng, and Peng Wang. A 3D photothermal structure toward improved energy efficiency in solar steam generation. *Joule*, 2(6):1171–1186, 2018.

[29] Juan Wang, Yangyang Li, Lin Deng, Nini Wei, Yakui Weng, Shuai Dong, Dianpeng Qi, Jun Qiu, Xiaodong Chen, and Tom Wu. High-performance photothermal conversion of narrow-bandgap Ti_2O_3 nanoparticles. *Advanced Materials*, 29(3):1603730, 2017.

[30] Xingyi Zhou, Fei Zhao, Youhong Guo, Brian Rosenberger, and Guihua Yu. Architecting highly hydratable polymer networks to tune the water state for solar water purification. *Science Advances*, 5(6):eaaw5484, 2019.

[31] Linfan Cui, Panpan Zhang, Yukun Xiao, Yuan Liang, Hanxue Liang, Zhihua Cheng, and Liangti Qu. High rate production of clean water based on the combined photo-electro-thermal effect of graphene architecture. *Advanced Materials*, 30(22):1706805, 2018.

[32] Anton Kuzyk, Robert Schreiber, Hui Zhang, Alexander O Govorov, Tim Liedl, and Na Liu. Reconfigurable 3D plasmonic metamolecules. *Nature materials*, 13(9):862–866, 2014.

[33] Na Liu, Martin Mesch, Thomas Weiss, Mario Hentschel, and Harald Giessen. Infrared perfect absorber and its application as plasmonic sensor. *Nano Letters*, 10(7):2342–2348, 2010.

[34] Yanming Liu, Shengtao Yu, Rui Feng, Antoine Bernard, Yang Liu, Yao Zhang, Haoze Duan, Wen Shang, Peng Tao, Chengyi Song, et al. A bioinspired, reusable, paper-based system for high-performance large-scale evaporation. *Advanced Materials*, 27(17):2768–2774, 2015.

[35] Oara Neumann, Alexander S Urban, Jared Day, Surbhi Lal, Peter Nordlander, and Naomi J Halas. Solar vapor generation enabled by nanoparticles. *ACS Nano*, 7(1):42–49, 2013.

[36] Kyuyoung Bae, Gumin Kang, Suehyun K Cho, Wounjhang Park, Kyoungsik Kim, and Willie J Padilla. Flexible thin-film black gold membranes with ultrabroadband plasmonic nanofocusing for efficient solar vapour generation. *Nature Communications*, 6(1):10103, 2015.

[37] Mark W Knight, Nicholas S King, Lifei Liu, Henry O Everitt, Peter Nordlander, and Naomi J Halas. Aluminum for plasmonics. *ACS Nano*, 8(1): 834–840, 2014.

[38] Christoph Langhammer, Markus Schwind, Bengt Kasemo, and Igor Zoric. Localized surface plasmon resonances in aluminum nanodisks. *Nano Letters*, 8(5):1461–1471, 2008.

[39] Emil Prodan, Corey Radloff, Naomi J Halas, and Peter Nordlander. A hybridization model for the plasmon response of complex nanostructures. *Science*, 302(5644):419–422, 2003.

[40] Jeffrey B Chou, Yi Xiang Yeng, Yoonkyung E Lee, Andrej Lenert, Veronika Rinnerbauer, Ivan Celanovic, Marin Soljačić, Nicholas X Fang, Evelyn N Wang, and Sang-Gook Kim. Enabling ideal selective solar absorption with 2d metallic dielectric photonic crystals. *Advanced Materials*, 26(47):8041–8045, 2014.

[41] Jianfeng Huang, Changxu Liu, Yihan Zhu, Silvia Masala, Erkki Alarousu, Yu Han, and Andrea Fratalocchi. Harnessing structural darkness in the visible and infrared wavelengths for a new source of light. *Nature Nanotechnology*, 11(1):60–66, 2016.

[42] Guohua Liu, Jinliang Xu, and Kaiying Wang. Solar water evaporation by black photothermal sheets. *Nano Energy*, 41:269–284, 2017.

[43] Fei Xiang, Xuhong Zhao, Jian Yang, Ning Li, Wenxiao Gong, Yizhen Liu, Arturo Burguete-Lopez, Yulan Li, Xiaobin Niu, and Andrea Fratalocchi. Enhanced selectivity in the electroproduction of H_2O_2 via F/S dual-doping in metal-free nanofibers. *Advanced Materials*, 35(7):2208533, 2023.

[44] Kun Jiang, Samira Siahrostami, Tingting Zheng, Yongfeng Hu, Sooyeon Hwang, Eli Stavitski, Yande Peng, James Dynes, Mehash Gangisetty, Dong Su, et al. Isolated ni single atoms in graphene nanosheets for high-performance CO_2 reduction. *Energy & Environmental Science*, 11(4):893–903, 2018.

[45] Ning Xu, Xiaozhen Hu, Weichao Xu, Xiuqiang Li, Lin Zhou, Shining Zhu, and Jia Zhu. Mushrooms as efficient solar steam-generation devices. *Advanced Materials*, 29(28):1606762, 2017.

[46] Jianhua Zhou, Yufei Gu, Pengfei Liu, Pengfei Wang, Lei Miao, Jing Liu, Anyun Wei, Xiaojiang Mu, Jinlei Li, and Jia Zhu. Development and evolution of the system structure for highly efficient solar steam generation from zero to three dimensions. *Advanced Functional Materials*, 29(50):1903255, 2019.

[47] Akanksha K Menon, Iwan Haechler, Sumanjeet Kaur, Sean Lubner, and Ravi S Prasher. Enhanced solar evaporation using a photo-thermal umbrella for wastewater management. *Nature Sustainability*, 3(2):144–151, 2020.

[48] George Ni, Gabriel Li, Svetlana V Boriskina, Hongxia Li, Weilin Yang, TieJun Zhang, and Gang Chen. Steam generation under one sun enabled by a floating structure with thermal concentration. *Nature Energy*, 1(9):1–7, 2016.

[49] Weichao Xu, Xiaozhen Hu, Shendong Zhuang, Yuxi Wang, Xiuqiang Li, Lin Zhou, Shining Zhu, and Jia Zhu. Flexible and salt resistant janus absorbers by electrospinning for stable and efficient solar desalination. *Advanced Energy Materials*, 8(14):1702884, 2018.

[50] Lianbin Zhang, Bo Tang, Jinbo Wu, Renyuan Li, and Peng Wang. Hydrophobic light-to-heat conversion membranes with self-healing ability for interfacial solar heating. *Advanced Materials*, 27(33):4889–4894, 2015.

[51] Hongqi Zou, Xiangtong Meng, Xin Zhao, and Jieshan Qiu. Hofmeister effect-enhanced hydration chemistry of hydrogel for high-efficiency solar-driven interfacial desalination. *Advanced Materials*, 35(5):2207262, 2023.

[52] Xiuqiang Li, Renxing Lin, George Ni, Ning Xu, Xiaozhen Hu, Bin Zhu, Guangxin Lv, Jinlei Li, Shining Zhu, and Jia Zhu. Three-dimensional artificial transpiration for efficient solar waste-water treatment. *National Science Review*, 5(1):70–77, 2018.

[53] Kaijie Yang, Tingting Pan, Saichao Dang, Qiaoqiang Gan, and Yu Han. Three-dimensional open architecture enabling salt-rejection solar evaporators with boosted water production efficiency. *Nature Communications*, 13(1):6653, 2022.

[54] Houze Yao, Panpan Zhang, Ce Yang, Qihua Liao, Xuanzhang Hao, Yaxin Huang, Miao Zhang, Xianbao Wang, Tengyu Lin, Huhu Cheng, et al. Janus-interface engineering boosting solar steam towards high-efficiency water collection. *Energy & Environmental Science*, 14(10):5330–5338, 2021.

[55] Xuan Wu, Yida Wang, Pan Wu, Jingyuan Zhao, Yi Lu, Xiaofei Yang, and Haolan Xu. Dual-zone photothermal evaporator for antisalt accumulation and highly efficient solar steam generation. *Advanced Functional Materials*, 31(34):2102618, 2021.

[56] Zhenxing Wang, Jie Gao, Jiajing Zhou, Jingwen Gong, Longwen Shang, Haobin Ye, Fang He, Shaoqin Peng, Zhixing Lin, Yuexiang Li, et al. Engineering metal–phenolic networks for solar desalination with directional salt crystallization. *Advanced Materials*, 35(1):2209015, 2023.

[57] George Ni, Seyed Hadi Zandavi, Seyyed Morteza Javid, Svetlana V Boriskina, Thomas A Cooper, and Gang Chen. A salt-rejecting floating solar still for low-cost desalination. *Energy & Environmental Science*, 11(6):1510–1519, 2018.

[58] Hardik K Jani and Kalpesh V Modi. Experimental performance evaluation of single basin dual slope solar still with circular and square cross-sectional hollow fins. *Solar Energy*, 179:186–194, 2019.

[59] Xiuqiang Li, Jinlei Li, Jinyou Lu, Ning Xu, Chuanlu Chen, Xinzhe Min, Bin Zhu, Hongxia Li, Lin Zhou, Shining Zhu, et al. Enhancement of interfacial solar vapor generation by environmental energy. *Joule*, 2(7):1331–1338, 2018.

[60] Xiuqiang Li, Xinzhe Min, Jinlei Li, Ning Xu, Pengchen Zhu, Bin Zhu, Shining Zhu, and Jia Zhu. Storage and recycling of interfacial solar steam enthalpy. *Joule*, 2(11):2477–2484, 2018.

[61] Hadi Ghasemi, George Ni, Amy Marie Marconnet, James Loomis, Selcuk Yerci, Nenad Miljkovic, and Gang Chen. Solar steam generation by heat localization. *Nature Communications*, 5(1):4449, 2014.

CHAPTER 11

Tailoring Transmission and Emission for Photonic Thermoregulating Textiles

Muluneh G. Abebe, Alice De Corte, and Bjorn Maes

University of Mons, Place du Parc, Belgium

11.1 INTRODUCTION

As humanity starts to experience the consequences of climate change, natural disasters due to extreme weather conditions have become frequent events. As far as humankind is concerned, global warming imposes a severe existential threat [1]. The imbalance between increasing energy consumption and clean energy production stands out. The rapid switch to renewable energy sources is difficult, and solutions to significantly decrease energy consumption are crucial. Surprisingly, more than half of our energy consumption goes to the heating and cooling of large, mostly empty spaces in residential and commercial buildings to keep the occupants in thermal comfort, while there is no difference in energy consumption between a few or many occupants [2]. Thus, a large amount of energy is wasted. Therefore, it is essential to reduce the heating and cooling consumption using personal thermal management, which creates a localized thermal regulation, with solutions such as heated chairs, leg warmers, directed heating, and smart fabrics. Controlling the radiative heat transfer for personal thermal-management technologies has gained much attention

DOI: 10.1201/9781003409090-11

due to its universality and high tunability, leading to photonic-engineered textiles. At a normal skin temperature of 34 °C, the human body loses its metabolically generated heat to a large extent by emitting infrared (IR) radiation centered near 10 μm. Specifically, in an indoor setting such as an office, more than 50% of the heat loss is attributed to IR radiation [3]. Therefore, with proper IR photon management, one can tailor and design passive temperature-regulating textiles.

To date, various photonic platforms and geometries have been investigated and utilized for temperature-regulating textiles. Generally, one distinguishes two approaches: modulating the transmission through the fabric or adjusting the emission from the fabric surface. Based on these concepts, the scientific community designed several state-of-the-art dual-mode (for both heating and cooling) fabrics with active (requires energy) or passive (requires no energy) switching capabilities [4]. For example, Leung et al. proposed a fabric that modulates the transmittance to deliver a heating and cooling performance [5]. Zhang et al. reported carbon nanotube-coated bimorph fibers to dynamically modulate the IR emissivity, providing both cooling and heating by adapting to the relative humidity [6]. Another interesting design based on transmittance modulation was presented with a bi-layer fabric from a combination of polyethylene terephthalate and moisture-responsive yarns [7]. In previous works, we demonstrated dynamic transmittance-modulating designs based on metallic and dielectric particles for dual-mode temperature regulation [8–11].

Here, following the design directions discussed above (i.e., transmission or emission modulation), we discuss two fabric designs: a Dynamic Transmittance Switch Textile (DTST), which modulates transmission, and a Janus-yarn fabric, which modulates emission. Both designs can deliver a dual-mode (cooling and heating) thermoregulating functionality by capitalizing on various thermal and photonic properties, with a switching capability via a shape memory polymer (dynamic) or via flipping of the fabric (static). This provides an opportunity to thermoregulate over a large range of ambient temperatures. This chapter combines and details the results from the work in [12, 13].

In Section 11.2 we present the proposed designs. Section 11.3 discusses the modeling methods, including an electromagnetic and a thermal analysis. In Section 11.4 we detail the results and the design performance. Section 11.5 concludes this chapter.

11.2 DESIGNS

In general, traditional textiles function in one mode, and can only be used comfortably in a limited window with respect to temperature and humidity. Therefore, one must adapt the clothing to different conditions to remain comfortable. For instance, in cold weather the body temperature is higher than the ambient, leading to radiative heat loss from the body. Thus, one can use a warm winter cloth that blocks heat loss from the body. On the other hand, in hot conditions one needs summer clothes that allow the body to relieve heat stress. However, it is impossible to use a single fabric comfortably in both conditions, which restricts their performance: there is no adaptation, with only a fixed function by design. Predominantly, traditional textiles possess high IR absorption/emission, combined with a low to moderate transmission, according to the material and the thickness. The traditional textile mainly controls conduction between the human body and the ambient, leaving the radiative channel largely unexploited. In contrast, our designs provide dual-mode functionality with both dynamic (DTST) and static (Janus-yarn) switching capability.

The DTST (Figure 11.1a) design incorporates fibers coated with a conductive material, which are arranged in an array, in a suitable combination with a shape-memory polymer [14]. The concept builds on material properties (i.e., scattering properties of the metallic coating, thermal and mechanical properties of the polymer actuator) and the geometry (i.e., shape and arrangement of the fibers). The dual-mode behavior is as follows: When the ambient temperature drops, the polymer beads shrink, thus decreasing the separation distance between two consecutive fibers. Consequently, the IR transmissivity will decrease, providing the desired heating function (Figure 11.1a(i)). Conversely, when the temperature rises, the polymer beads expand, thus increasing the inter-fiber distance. This results in a new configuration, with an increased IR transmissivity providing the cooling functionality (Figure 11.1a(ii)).

A metal is used as a coating material due to its high reflectivity in the wavelength range of human body emission. In contrast, with dielectric fibers (without metal) below an optimum diameter, the IR transmission would rapidly increase. However, due to conductivity, metal-coated fibers act like antennae, which strongly interact with IR radiation, even if the fiber diameter is much smaller than the wavelength. In this way, a strong IR transmission control is possible, with efficient thermal body regulation,

(a) DTST fabric (dynamic)

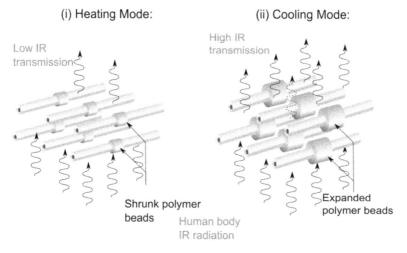

(i) Heating Mode:

(ii) Cooling Mode:

Low IR transmission

High IR transmission

Shrunk polymer beads

Expanded polymer beads

Human body IR radiation

(b) Janus-yarn fabric (static)

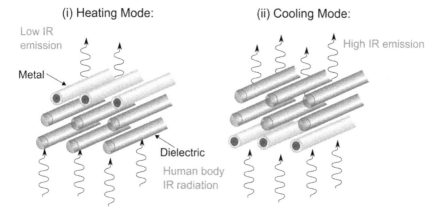

(i) Heating Mode:

(ii) Cooling Mode:

Low IR emission

High IR emission

Metal

Dielectric

Human body IR radiation

FIGURE 11.1   Working principle: (a) DTST fabric, (b) Janus-yarn fabric.

as shown later. Metals such as silver, gold, copper, and aluminum can all be used as a coating on dielectric-core fibers, which can be composed of natural or synthetic textile materials such as cotton and polyester.

The Janus-yarn fabric (Figure 11.1b) employs the Stefan-Boltzmann radiation emission law, stating that the power radiated is proportional to the emissivity ε of the surface and the fourth power of its temperature: $P = \varepsilon \sigma T^4$, where σ is the Stefan-Boltzmann constant. Therefore, engineering the surface emissivity controls the amount of radiation transferred to the ambient. Furthermore, according to Kirchhoff's law, at equilibrium, the absorptance equals the emissivity; so enhancing and suppressing absorption in the fabric will adjust the emissivity. Therefore, when the low emissive layer of the fabric – highly reflecting metal-coated fibers – faces the ambient, the surface acts as infrared radiative insulation, creating a heating effect in a cold environment (Figure 11.1b(i)). On the other hand, when it is hot, flipping the same fabric exposes the high emissivity side – dielectric fibers – to the ambient, acting as an IR radiator, thus delivering a cooling function (Figure 11.1b(ii)). Furthermore, because the fabric is constituted out of yarns, which are bundles of fibers, it provides the required air permeability and water-vapor transmission for standard thermal comfort.

The asymmetric emissivity of the fabric is achieved from the structure of the Janus-yarn, using two different materials: highly absorbing/emitting dielectric fibers, and highly reflecting non-absorbing/emitting metallic fibers. Both dielectric and metallic fibers are staggered in a specific photonic geometry to provide optimal functionality. Consequently, for the heating (resp. cooling) mode, the metal (resp. dielectric) side of the yarn faces the ambient.

11.3 MODELING METHODS

The overall performance of the proposed designs is examined by modeling the heat transfer from the skin to the ambient. The main objective is to determine the maximum and minimum ambient temperatures that maintain the wearer's thermal comfort: the setpoint temperature. To determine this temperature, first we utilize an electromagnetic simulation to retrieve spectral and radiative parameters. Subsequently, we feed these characteristics into the thermal model and analyze the fabric performance.

11.3.1 Electromagnetic Modeling

To study the IR response we employ the finite-element method to calculate rigorous solutions of Maxwell's equations, using commercial software (COMSOL Multiphysics). Since the fibers are considered infinitely long, the two-dimensional (2D) geometries consist of fibers arranged in hexagonal geometry (Figure 11.2a, b). For DTST, the geometry is constituted from metallic fibers with radius r and center-to-center distance d (Figure 11.2a). For Janus-yarn, there are metallic fibers (radius r_1) on one side, and dielectric fibers (radius r_2) on the other side (Figure 11.2b), and with center-to-center distance d.

In the relevant IR range, many metals can be approximated as perfect electric conductors, with an infinite conductivity. Thus, we implement the

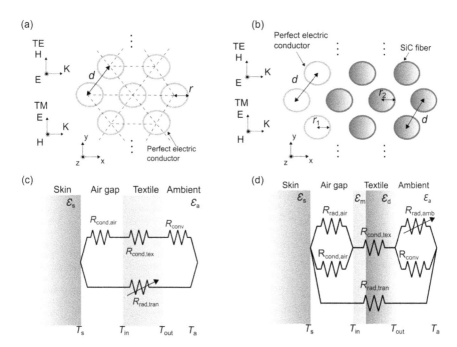

FIGURE 11.2 The simulation model, with the polarization and incidence direction for: (a) DTST, (b) Janus-yarn fabric. Patterns repeat in the y-direction. Thermal circuit analog for (c) DTST and (d) Janus-yarn fabric

Perfect Electric Conductor (PEC) boundary condition, which leads to a complete reflection from these surfaces without any losses. For the Janus-yarn design, due to its strong IR absorption in the relevant wavelength region (4–25 µm), silicon carbide (SiC) is used as the dielectric material, with refractive indices derived from [15]. However, other materials with significant IR absorption (such as cellulose and polydimethylsiloxane) could be employed accordingly.

In the simulations a perpendicular plane wave source impinges on the structures. The IR spectral parameters strongly depend on the polarization. Therefore, it is of crucial importance to take into account both a transverse electric (TE, one electric field component out-of-plane) and transverse magnetic (TM, one magnetic field component out-of-plane) case. For the DTST design, excitation from one side of the structure is enough, while for the Janus-yarn, incidence from both sides is required: metallic (left in Figure 11.2b) and dielectric side (right in Figure 11.2b). Floquet periodic boundary conditions are used on the top and bottom to represent an infinite repetition in the vertical direction. The various diffraction orders for smaller wavelengths are computed with Port conditions on the right and left. Due to computation time, the geometric optimization is done using 2D simulations. However, full three-dimensional (3D) simulations are deployed to validate the 2D results, and for the best designs.

For the thermal model, one needs the total radiative quantities: transmittance (τ), absorptance ($\alpha = \varepsilon$), and reflectance (ρ). These are the spectrally integrated transmission T, reflection R, and absorption A, weighted with the human body emissivity, for which we employ the Planck distribution ϕ_{bb} with a skin temperature of 34 °C:

$$\tau/\alpha/\rho = \frac{\int_{\lambda_1}^{\lambda_2} T/A/R(\lambda)\phi_{bb}(\lambda)d\lambda}{\int_{\lambda_1}^{\lambda_2} \phi_{bb}(\lambda)d\lambda}, \tag{11.1}$$

with the wavelength range from $\lambda_1 = 4$ μm to $\lambda_2 = 25$ μm, containing most of the human body radiation.

11.3.2 Thermal Modeling

The cooling and heating performance is assessed with the ambient set-point temperature, which is determined from a heat transfer analysis using a circuit model. We treat thermal dissipation from the human body to the ambient as a one-dimensional (1D) steady-state heat transfer problem in dry conditions, hence ignoring mass transfer. We model the fabric designs separately since DTST is a transparent (non-absorbing/non-emitting) layer, while the Janus-yarn is semitransparent (absorbing and emitting).

For transparent (non-absorbing/non-emitting) fabric layers, such as DTST, the radiative transfer in the air gap, in the fabric and in front of the fabric is the same. Conduction in the air gap, conduction in the fabric and convection from the fabric surface are the other contributions. For a semitransparent (absorbing and emitting) fabric layer, such as Janus-yarn, radiative and conductive heat transfer constitute the exchange through the air gap, while the heat dissipation from the fabric to the ambient is via radiation and convection. Another conductive exchange contributes to the heat transfer in the textile. Furthermore, radiation exchange occurs between the skin and the ambient with transmission through the fabric layer. When the transmission through the fabric is low, one can assume an opaque layer.

A thermal circuit model is constructed for both cases involving the conductivity of the textile and its radiative properties, obtained from the electromagnetic simulations, as well as properties of the air gap and ambient, see Figure 11.2c, d. The R symbols represent the thermal resistances, and the corresponding subscripts denote radiation in air, transmitted through fabric, and the ambient, conduction in air and the textile, and convection, respectively. Furthermore, we use the inside and outside temperatures T_{in} and T_{out}, the skin temperature T_s, and the ambient temperature T_a. The requirement for a wearer's thermal comfort is the balance between metabolic heat generation and total heat loss in dry conditions. The total heat loss is controlled by the total thermal resistance between skin and ambient. Due to the change in radiative properties of the proposed fabric designs, the total thermal resistance varies between cooling and heating mode.

The heat transfer model assumes a constant body heat generation of $Q = 70$ W/m^2 corresponding to a sedentary individual with a skin of 34 °C. A typical air gapof 1 mm is taken for the microclimate thickness. Furthermore, the thermal conductivity of air is $k_{air} = 0.03$ Wm^{-2}K^{-1}. The thermal conductivity (k_{eff}) of the fabric is taken as an effective value of the contributing components (metallic fibers and air in DTST, and metallic, dielectric fibers and air in Janus-yarn). The natural convection heat transfer coefficient is $h = 3$ Wm^{-1}K^{-1}. The emissivity of the skin is approximated as a gray body with $\varepsilon_s = 0.98$, the emissivity of ambient as a black body is $\varepsilon_a = 1$. The thickness of the DTST follows $4(d + r)$ and the Janus-yarn fabric is ~20 μm. We assume that the convective heat transfer in the air gap is negligible for both designs. This is due to the small Rayleigh number, which stems from the relatively small air gap thickness [16].

11.4 RESULTS AND DISCUSSION

Herein, we discuss in detail the electromagnetic response for both TE and TM polarizations for each design. Furthermore, we evaluate the overall efficiency with respect to the user's comfort based on the heat transfer analysis.

11.4.1 Electromagnetic Response of DTST Fabric

The TE transmission spectra are very distinctive and provide good opportunities for thermal tuning. They are shown in Figure 11.3 for selected parameters: each panel fixes f and shows spectra with varying d. The spectra have specific wavelength regions or bands due to several photonic effects. One notices a similarity between the curves, with varying band positions along the wavelength axis (shifting to right/left) and band magnitudes (higher/lower transmittance).

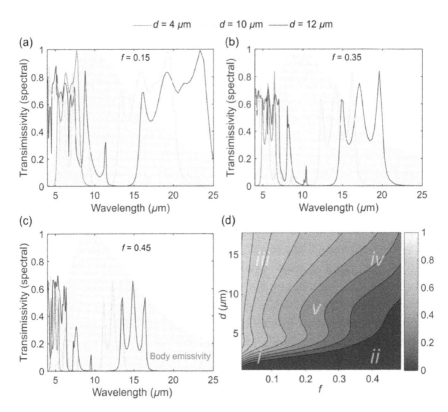

FIGURE 11.3 TE transmittance as a function of wavelength for $d = 4$, 10, and 12 μm and (a) $f = 0.15$, (b) $f = 0.35$, (c) $f = 0.45$. (d) Map showing τ as a function of f and d.

For a clear example, we examine the curve for $d = 10$ μm in Figure 11.3b ($f = 0.35$, line). For larger wavelengths, above a cut-off wavelength of about 17 μm, the spectrum exhibits a wide (semi-infinite) stopband with low transmission, called the plasmonic gap, leading to a complete blockage of thermal radiation transmission [17]. Interestingly, this gap does not originate from the geometry of the design, but rather from the physical property of the perfectly conducting scatterers.

Next to the plasmonic gap, there is a first transmission band, extending from 12 to 17 μm, with a center wavelength following the Bragg condition, similar to dielectric photonic crystal bandgaps. This transmission band presents three resonance peaks, similar to Fabry-Pérot cavity modes of finite photonic crystals or multilayers with only a few layers [18].

In addition to the plasmonic gap, we observe a structural bandgap for wavelengths just below the first transmission band, which originates from the geometric design, and extends from 9 to 12 μm, again blocking the transmission of thermal radiation. Finally, below the structural bandgap, there is a continuous second transmission band below 9 μm, with a complicated set of resonances.

Now we can assess the impact of d and f on these ranges. When d increases (decreases), the spectra shift to longer (shorter) wavelengths. For example, for $f = 0.35$ (Figure 11.3b) when increasing d from 10 to 12 μm, the spectrum shifts to the right: plasmonic gap (from > 17 μm to > 22 μm), first transmission band (from 12-17 μm to 14-21 μm), structural bandgap (from 9-12 μm to 11-14 μm), and second transmission band (from < 9 μm to < 11 μm).

For the effect of the fill factor f we can compare e.g., the three spectra for $d = 10$ μm (Figure 11.3a–c for $f = 0.15, 0.35$, and 0.45). Overall, the magnitude of the transmission bands decreases when f increases, as we indeed expect more scattering. Furthermore, if we examine the first transmission band, the large wavelength edge shifts much more than the small wavelength edge. Consequently, for increasing f, the first transmission band becomes noticeably narrower. The structural bandgap, however, remains fairly constant in width.

Thus, variation of d and f has significant consequences, especially concerning band shifting. This effect is imperative for the dynamic fabric, as the human body emissivity is fixed (background in Figure 11.3a–c), while the textile transmission can switch by tuning d and f. Now, we need to find optimal parameter values that allow the fabric to operate efficiently in two different modes (cooling and heating), for which we use the blackbody-weighted integrated transmittance τ (see Equation 11.1) as a figure-of-merit.

Figure 11.3d maps τ as a function of d and f, where we can distinguish five regions: Region (i), bottom-left corner where both f and d are very small ($f < 0.1$, $d < 2$ μm), (ii), bottom-right corner, large f and relatively small d ($f > 0.4$ and $d < 2$ μm), (iii), top-left corner, small f and large d ($f < 0.1$ and $d > 10$ μm), (iv), top-right corner where both f and d are very large, (v), the middle, where d is from 2 to 10 μm and the whole range of $f = 0.02$ to 0.45. We now discuss these regions, with an eye to dynamic textile applications.

In region (i), τ is close to zero, and this implies that when both f and d are small, the design will reflect most of the thermal radiation back to the human body. As a result, the DTST operates in heating mode. The very low τ in this region is because the plasmonic gap shifts to short wavelengths, as far as below 5 μm covering most of the human body emissivity curve. So the first transmission band has almost no overlap with the human emissivity, resulting in a very low τ.

In region (ii), τ becomes even smaller. With increasing f, the transmission band becomes weaker, and shifts even further to smaller wavelengths, leading to even less overlap. Therefore, the fabric can operate in the heating mode in this region as well, profiting from the plasmonic gap.

In region (iii), τ is above 0.7 due to a large d and small f, which fosters the dominance of the second transmission band. In this case, both the plasmonic gap and the structural bandgap are shifted towards larger wavelengths, beyond the main human body emissivity range. Therefore, thermal radiation transmission from the human body to the environment increases; thus, this region can facilitate a cooling mode.

Region (iv) presents a moderate τ, in the range of 0.4 to 0.5. In this case, the important second transmission band due to increased d, is counterbalanced by the decrease of the magnitude of this specific band due to an increase in f. In addition to this, the structural bandgap broadens due to f. Moreover, the plasmonic gap shifts to shorter wavelengths, decreasing τ (see $d = 12$ μm in Figure 11.3a–c).

In region (v) there is a bump around $d = 4$–5 μm that exists for a wide range of f, and then stretches and flattens for higher f. Interestingly, the reason behind this bump is the dominance of the first transmission band under the emissivity curve for $d = 4$ μm (in Figure 11.3a–c). More specifically, for very small d the first transmission band has not entered under the human body emissivity curve, yet gradually with increasing d it shifts to the right and overlaps. Around $d = 4$ μm the majority of this band overlaps, leading to a higher τ bump. Furthermore, since the first transmission band shrinks with increasing f, the bump flattens for larger f. One has to keep in mind that overall, increasing f decreases τ. The dip around $d = 6$ μm, right after the bump, is due to the structural bandgap sliding under the emissivity curve. Upon further increasing f the structural bandgap shifts and is replaced by the second transmission band.

Figure 11.3d allows to identify a range of parameters f and d that are suitable for a dynamic textile. Each τ point on the map corresponds to a certain d, f, and r, as $f = 2r/d$. Thus, for most variations in the map, the radius of the fibers r changes, which is difficult in a practical realization. However, one can fix r and assess the behavior of τ as a function of f and d. This is done by extracting τ from the map (Figure 11.3d) for a fixed r, which gives a hyperbola when plotted as a function of f and d, see Figure 13.4a. The color bar

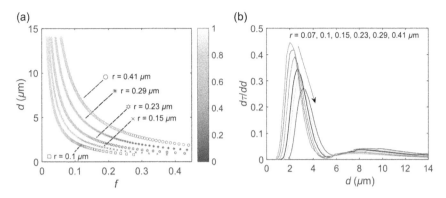

FIGURE 11.4 (a) τ as a function of f and d for constant radii (r = 0.1, 0.15, 0.23, 0.29, and 0.41 μm). (b) $d\tau/dd$ as a function of d for various r values for TE polarization.

illustrates the magnitude of τ, so the change of τ is visible along the hyperbolas, and this is the effect we can exploit for a dynamic textile.

Furthermore, Figure 11.4a shows that τ is large for small f and large d, and vice versa. One also notices the curvature of the hyperbolas for different r's: smaller r corresponds to a larger curvature. The hyperbolas with larger curvature have a sharper color change (larger gradient) for specific values of f and d. This sharp color change is important for finding a sweet spot for switching from one mode to the other in a fabric. More specifically, a drastic change in τ due to small changes in geometric parameters leads to a highly dynamic transmittance.

Considering this, we introduce as figure-of-merit the derivative of τ with respect to d, see Figure 11.4b. With increasing r, $d\tau/dd$ decreases and shifts to larger d. For $r < 0.1$ μm there is a very large value of $d\tau/dd$ (0.45–0.55 μm^{-1}), decreasing upon increasing r. This is because a smaller r corresponds to a smaller f, which results in a larger τ and $d\tau/dd$. Now we can determine a suitable r for efficient operation: a very small r gives the largest and best $d\tau/dd$, and accordingly the best switching efficiency. The practical limit for the size of mono-filament diameter in current textile technology can go to the sub-micrometer scale. For example, for a carbon-based nanofiber the reported diameter is around 200 nm [19]. If this is the fabrication limit, $r < 0.1$ μm is not practical. As a result, we use $r \geq 0.1$ μm, which corresponds to $d\tau/dd \leq 0.4$ μm^{-1} for the proposed fabric.

After TE, we move on to the TM polarization (spectra in Figure 11.5a–c). For longer wavelengths, unlike TE, there is no plasmonic gap. Instead, there is a transmission band followed by a structural bandgap. Furthermore, this

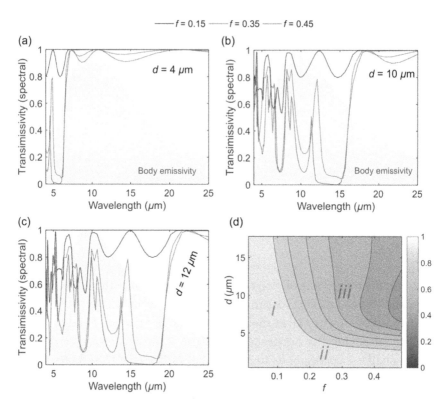

FIGURE 11.5 TM transmittance spectra for $f = 0.15$, 0.35, and 0.45 μm, and (a) $d = 4$ μm, (b) $d = 10$ μm, (c) $d = 12$ μm. (d) Spectrally integrated transmittance τ for TM as a function of f and d.

structural bandgap has a strong dependence on f, and for larger f, this bandgap becomes quite deep (see $f = 0.45$ in Figure 11.5a–c). Meanwhile, for smaller f, the structural bandgap is weak and insignificant (see $f = 0.15$ in Figure 11.5a–c). Furthermore, the spectra shift to the right (left) with increasing (decreasing) d.

Figure 11.5d shows the map of τ as a function of f and d. We can now divide this map in three parts: region (i) where $f < 0.1$, region (ii) where $f > 0.1$ and $d < 4$ μm, and region (iii), where $f > 0.15$ and $d > 4$ μm. In regions (i) and (ii) τ is around 0.9, thus, close to a complete thermal transmission, which is in contrast with region (i) for TE (see Figure 11.3d), but analogous to region (iii) for TE (very small f and very large d in Figure 11.3d). In region (iii) τ starts to decrease gradually.

Thermal radiation emitted by the human body is a largely incoherent, unpolarized electromagnetic wave, consisting of both TE and TM components.

So far, we performed calculations for TE and TM independently, to identify the various photonic effects as a function of the geometric parameters. Here, we further assess these findings for flexible textiles. For example, we explore various fiber configurations that could lead to a dynamic textile.

The TE calculations show a fortunate result, mainly due to the plasmonic gap and its dynamic response. Contrarily, since the plasmonic gap does not exist for TM, it is not possible to utilize the same effect for both polarizations at the same time. However, to benefit both polarizations, a crossed-array configuration provides a solution (Figure 11.6a). In the crossed case, the polarization component, which is TE for the first array, becomes TM for the second array (and vice versa). Thus, if we use the edge of the plasmonic gap to modulate the transmission of the TE component via the first array, the transmission through the second array will be quasi-complete. Similarly, for the TM component, the transmission through the first array will be almost total, and the second array can be used to modulate the transmission (as TM becomes TE for the second array). In other words, a polarization modulated in the first array will be largely transmitted through the second array (and vice versa). We justify this reasoning using a 3D simulation for the crossed array configuration (Figure 11.6b). The spectral transmittance for the full 3D simulation (crossed wires) is like the 2D simulation (non-crossed). The difference consists of several additional Fabry-Pérot peaks, which do not significantly influence the integrated figures-of-merit.

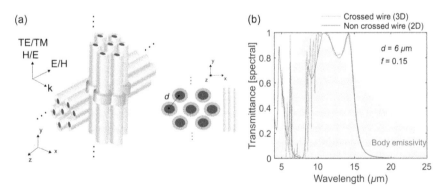

FIGURE 11.6 (a) Crossed-array configuration: (left) perspective and (right) planar view. (b) Transmittance comparison between crossed wire (3D simulation) and non-crossed wire (2D simulation, TE) configuration for $d = 6$ μm and $f = 0.15$.

Thus, the crossed array can fully control the transmission of thermal radiation from the human body. It is also worthwhile to note that such a crossed nature is standard in textile production, especially for woven fabrics.

11.4.2 Electromagnetic Response of Janus-Yarn Fabric

For a given geometry ($r_1 = 0.6\ \mu m$, $r_2 = 1.125\ \mu m$, $d = 3\ \mu m$), the IR response from the metallic side (A, T, R) for TE (solid lines) and TM (dashed) is displayed in Figure 11.7a. For TE, R and T show a plasmonic gap stationed above a cutoff around $\lambda = 4\ \mu m$, a fundamental metallic effect (similar to DTST) that blocks the TE radiation from transmitting to the dielectric

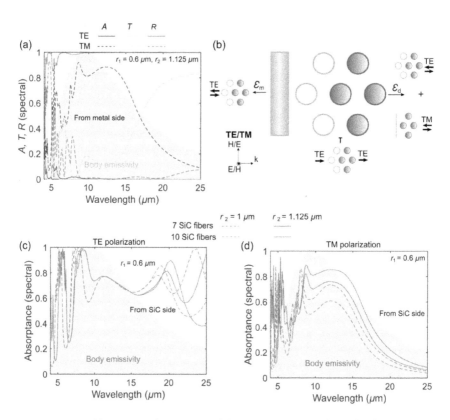

FIGURE 11.7 (a) TE/TM absorptance (A), transmittance (T), and reflectance (R) of Janus-yarn ($r_1 = 0.6\ \mu m$, $r_2 = 1.125\ \mu m$, $d = 3\ \mu m$) illuminated from metallic side. (b) Schematic of the 2D radiative parameters to approximate the crossed fiber behavior. (d) TE and (e) TM absorptance (A) for fixed r_1 (0.6 μm) while varying r_2 (1 and 1.125 μm) and number of SiC fibers (7 and 10), incidence from dielectric side.

fibers and through the yarn. As a result, there is a close to zero overall TE absorption leading to a very low emission from the yarn's metallic side.

However, unlike the TE case, there is no plasmonic gap for TM polarization (similar to DTST). Therefore, TM is transmitted through the metallic fibers, and a substantial amount is absorbed in the dielectric fibers or reaches the ambient. Unfortunately, this TM absorption and transmission increases the averaged, unpolarized characteristics, which would negatively affect the heating mode operation. However, a crossed configuration of the metallic fibers will benefit both polarizations, which will drastically cut the unpolarized transmission and absorption.

Here the crossed metal wire configuration is similar to the DTST situation and works as follows: the TE orientation for the first array behaves as TM for the second (90° rotated) array, and vice versa (Figure 11.7b). Subsequently, there is always an array with an electric field parallel to the wires, leading to a strong reflection on one of the arrays due to the plasmonic gap. Accordingly, the total emissivity from the metal side with crossed wires, is very close to the spectrally integrated value for TE in the 2D situation (without crossed wires, which is validated with 3D simulations), about $\varepsilon_m = 0.017$, guaranteeing a very low emissivity from the metallic side and allowing a heating mode.

Most of TE and TM impinging from the dielectric side enters the Janus-yarn and is substantially absorbed by the dielectric fibers (Figure 11.7c, d). Radiation that passes the dielectric fibers in TE will be reflected by the metallic fibers due to the plasmonic gap (acting as an IR mirror), enhancing the absorption path length and the emission. On the other hand, the TM radiation that manages to pass the dielectric will largely pass the metallic fibers, as the plasmonic gap is absent, which negatively affects the emission. Note that this can again be solved by using the crossed wire configuration.

In cooling mode, the main design goal is to achieve a high emissivity from the dielectric side; hence, one can optimize the various geometric parameters to increase the absorption: fiber size, separation distance, number of fibers and so on. To illustrate, we present the absorptance for a fixed fiber separation distance ($d = 3$ μm) with varying fiber size ($r_2 = 1$ and 1.125 μm) and number of SiC fibers (7 and 10), see Figure 11.7c, d. The geometry with ten SiC fibers contains three additional fibers to the right of Figure 11.2b (adding another layer in the hexagonal arrangement).

Under the human body emissivity curve, for TE (Figure 11.7c), the absorption contrast between different fiber sizes, as well as between different fiber numbers is small. On the other hand, for TM (Figure 11.7d), there exists a noticeable contrast: increasing the size and the number of fibers gives a better absorption. For example, the best performing structure is the largest $r_2 = 1.125$ μm with ten fibers. While the least performing is $r_2 = 1$ μm with seven fibers.

Next, we vary the fiber separation (d = 2, 3, and 5 μm) for a fixed radius ($r_1 = 0.6$ μm and $r_2 = 1.125$ μm) and ten dielectric fibers, see Figure 11.8a for TE absorption incidence from the dielectric side. In general, the absorption is high; however, there are variations originating from various photonic characteristics. One distinguishes three regions: (i) Narrow peaks at shorter wavelengths (4–9 μm), (ii) a relatively constant high absorption plateau at medium wavelengths (10–15 μm), (iii) broader resonances at longer wavelengths (16–25 μm).

These regimes are interpreted via the field profiles (electric field norm) at specific peaks, see Figure 11.8b. Within region (i), panels (3)–(6) show

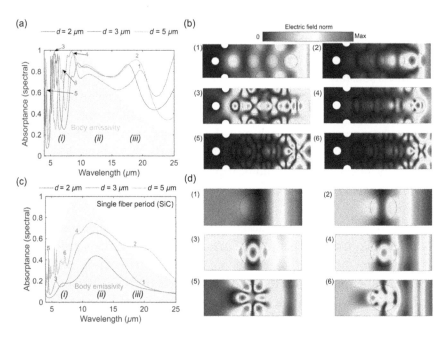

FIGURE 11.8 (a) TE absorptance of a yarn ($r_2 = 1.125$ μm) for separations d of 2, 3 and 5 μm, dielectric incidence. (b) Electric field norm at: (1) λ = 19.8 μm, (2) λ = 19 μm, (3) λ = 5.5 μm, (4) λ = 8.7 μm, (5) λ = 4.45 μm, (6) λ = 7.13 μm.

localized higher-order multipole type resonances, with e.g., quadrupoles in panel (3) ($d = 3$ µm, $\lambda = 5.5$ µm) and hexapoles in panel (5). This leads to multiple, narrowband increases of the local field and the absorption. Within region (iii), panel (1) ($d = 3$ µm, $\lambda = 19$ µm) shows globally a Fabry-Pérot type mode, with an envelope and multiple maxima. Panel (2) ($d = 3$ µm, $\lambda = 17.6$ µm) develops coupled localized dipolar modes. The latter modes are lower order, and thus more broadband. On the other hand, region (ii) is relatively featureless, leading to a plateau of high absorption, which can be understood via the large absorption of SiC in this range. The comparison with a single dielectric fiber period of the same size is studied extensively in [13]. Overall, one observes that the interplay of local and global field enhancements aids to achieve strong absorption/emission via the dielectric fibers.

With these results, an overall optimization using 2D simulations (with the crossed wire approach in mind) indicates the following geometric parameters: Three metallic fibers per period ($r_1 = 0.6$ µm, $d = 3$ µm) to sufficiently cut the transmission ($\tau = 0.0035$), leading to a low total emissivity ($\varepsilon_m = 0.017$) for heating performance. The latter combined with ten dielectric fibers ($r_2 = 1.125$ µm, $d = 3$ µm), delivering a large total emissivity ($\varepsilon_d = 0.72$) required for cooling.

To confirm the 2D results' accuracy, and the crossed-wire method, we employ 3D simulations with actual crossed metallic fibers (Figure 11.9a). The spectral response for unpolarized radiation (averaged polarization)

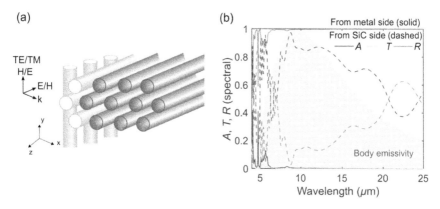

FIGURE 11.9 (a) The crossed fibers in 3D. (b) Unpolarized spectral absorptance, transmittance and reflectance of a Janus-yarn (crossed-wire 3D geometry) for the best parameters (10 SiC fibers, 3 metallic fibers, $r_1 = 0.6$ µm, $r_2 = 1.125$ µm, $d = 3$ µm), with incidence from the metal (dashed lines) and dielectric (solid lines) side.

for incidence from the metallic and dielectric side is displayed in Figure 11.9b. From the metallic side, a very low spectral absorptance and transmittance is achieved (Figure 11.9b), leading to a very low total emissivity ($\varepsilon_m = 0.017$) and transmittance ($\tau = 0.0039$). Therefore, crossing the metallic fibers has the expected, very useful effect, a plasmonic gap for both polarizations. From the dielectric side (dashed lines in Figure 11.9b), the absorptance is very large: above 0.7 from 8 to 15 μm and almost reaching 1 at a peak around 9 μm, resulting in a high total emissivity ($\varepsilon_d = 0.74$). Finally, the transmittance is very low from both sides, which is important for thermal properties.

11.4.3 Fabric Performance

From the thermal circuit analysis for crossed-fiber DTST, an expanded thermal comfort range is achievable, with lowest setpoint 9.5 °C, and highest setpoint 25.7 °C (Figure 11.10a). The lowest setpoint corresponds to τ close to zero, where $d = 1.2$ μm (expected maximum shrinking limit of polymer beads), and the plasmonic gap covers most of the human body emissivity curve; thus, the fabric is opaque to thermal radiation emitted by the human body. On the other hand, the highest setpoint achieved by DTST corresponds to $\tau = 0.9$, where $d = 14$ μm (expected maximum swelling limit of polymer beads), and the second transmission band is the dominant photonic effect under the human body emissivity curve; thus, the fabric is highly transparent to thermal radiation emitted by the human body. In addition, it is already possible to achieve a satisfactory highest

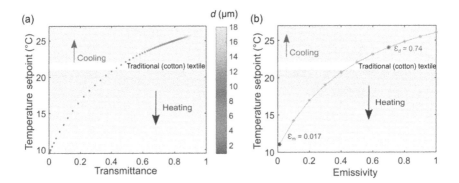

FIGURE 11.10   Ambient setpoint temperature: (a) For DTST as a function of τ and d in crossed configuration for $r = 0.1$ μm. (b) For Janus-yarn fabric as a function of the textile's IR outward emissivity. The small comfort range for traditional (cotton) textiles is also indicated.

setpoint temperature of ~24.5 °C with a limited swelling to only $d = 4$ μm, which corresponds to $\tau = 0.68$. Overall, the dynamic woven configuration provides a very wide ambient setpoint temperature window of ~16 °C.

The maximum, ideal upper setpoint temperature of an emissivity-modulating cooling fabric is 26 °C, which is achieved with the outer surface emissivity value of one, while the inside emission does not play a role any longer. The minimum, ideal lower setpoint temperature of an emissivity-modulating heating fabric is 11 °C, which is reached when outer surface emissivity is zero, and the inside emission again is not a factor. However, varying the outer emissivity while fixing the inside to one/zero (i.e., both give the same result) results in a wide setpoint window (Figure 11.10b). This demonstrates the vital importance of the outer surface emissivity of the Janus-yarn fabric to modulate between upper and lower setpoint temperatures to provide thermal comfort in a very low transmission situation. Based on this analogy, the proposed Janus-yarn fabric attains 24.4 °C (upper star on the curve) highest setpoint temperature in cooling mode (dielectric fibers facing outside, $\varepsilon_d = 0.74$), and 11.3 °C (lower star on the curve) lowest setpoint in heating mode (metallic fibers facing outside, $\varepsilon_m = 0.017$). This leads to a very large thermal temperature window of 13.1 °C.

To compare their performance with other existing technologies, for DTST and Janus-yarn fabric, the lowest setpoint is 4.8 °C and 3 °C lower than for the Mylar/space blanket (14.3 °C) [20], 4.4 °C and 2.6 °C lower than for the dual-mode textile (13.9 °C) [21], 5.2 °C and 3.4 °C lower than for the nanoporous metalized polyethylene textile (14.7 °C) in its cotton/polyethylene/Ag configuration [22]. Therefore, both designs' heating functionality is not only better than typical traditional winter fabrics, but also shows superiority over various advanced thermal management textiles. Furthermore, the highest setpoint is 2.9 °C and 1.6 °C larger than that of cotton (22.8 °C), 1.4 °C and 0.1 °C more than for the dual-mode textile (24.3 °C) [21]. In essence, our designs are promising and capable of preserving the thermal comfort of the wearer in a highly dynamic temperature situation.

11.5 CONCLUSION AND OUTLOOK

Thermal comfort is an important human sensation, with clothing as the main means to achieve it. In general, textiles control the heat transfer from the body to the ambient to ensure thermal balance and comfort. However, traditional textiles are insufficient, and external heating and cooling systems are used. Furthermore, the energy spent for (mostly empty) spaces of buildings becomes an enormous drain. Traditional textiles can only function in one mode and are usable in a limited temperature window. Therefore, one must adapt the clothing to different conditions: warming textiles for cold surroundings and cooling textiles in warm environments.

To address these problems, we propose fabric designs for personal thermal regulation based on photonic engineering. The designs possess dual-mode functionality (cooling and heating), one with a switching capability via shape memory polymers, and the other employing a static switch (i.e., fabric flipping). Extensive electromagnetic and thermal modeling is implemented to demonstrate the potential thermoregulating benefits.

The first design (DTST fabric) is constituted of metal-coated fibers and stimuli-responsive polymer beads, which enable a passive dynamic response. The design benefits from various IR photonic effects to strongly control the wide-band transmission ($\Delta\tau = 0.9$) of thermal radiation and provide a wide temperate setpoint window of ~16 °C. The Janus-yarn fabric exhibits a strongly asymmetric emissivity, due to a tailored choice of materials (metallic and dielectric fibers) and geometries, providing optimized exploitation of photonic effects, such as the plasmonic gap and localized resonances. The fabric switches from one mode to the other by mechanical flipping with an emissivity contrast of about 0.7 between the two surfaces, leading to a wide comfort window of 13.1 °C.

The envisioned designs can be fabricated by utilizing micro/nanofabrication techniques combined with standard textile manufacturing methods. These techniques can vary from the very advanced 3D printing and core-shell electrospinning to the basic extrusion and dying processes. In general, our designs provide an opportunity to thermoregulate over a large range of ambient temperatures. Subsequently, these approaches demonstrate a large potential for replacing heating and cooling systems, which allows a substantial energy saving.

The authors thank the INTERREG PHOTONITEX, FNRS postdoctoral, FNRS-FRIA doctoral, and ARC-21/25 UMONS2 projects. We also acknowledge Gilles Rosolen, Eric Khousakoun, Jeremy Odent, Jean-Marie Raquez, and Sylvain Desprez for their fruitful collaboration.

REFERENCES

1. S. Banholzer, J. Kossin, and S. Donner, "The impact of climate change on natural disasters," in *Reducing disaster: Early warning systems for climate change*, pp. 21–49, Springer, 2014.

2. A. Ghahramani, K. Zhang, K. Dutta, Z. Yang, and B. Becerik-Gerber, "Energy savings from temperature setpoints and deadband: Quantifying the influence of building and system properties on savings," *Applied Energy*, vol. 165, pp. 930–942, 2016.

3. J. D. Hardy and E. F. DuBois, "Regulation of heat loss from the human body," *Proceedings of the National Academy of Sciences of the United States of America*, vol. 23, no. 12, p. 624, 1937.

4. Y. Fang, G. Chen, M. Bick, and J. Chen, "Smart textiles for personalized thermoregulation," *Chemical Society Reviews*, vol. 50, pp. 9357–9374, 2021.

5. E. M. Leung, M. C. Escobar, G. T. Stiubianu, S. R. Jim, A. L. Vyatskikh, Z. Feng, N. Garner, P. Patel, K. L. Naughton, M. Follador, et al., "A dynamic thermoregulatory material inspired by squid skin," *Nature Communications*, vol. 10, no. 1, pp. 1–10, 2019.

6. X. A. Zhang, S. Yu, B. Xu, M. Li, Z. Peng, Y. Wang, S. Deng, X. Wu, Z. Wu, M. Ouyang, et al., "Dynamic gating of infrared radiation in a textile," *Science*, vol. 363, no. 6427, pp. 619–623, 2019.

7. Fu, Z. Yang, Y. Pei, Y. Wang, B. Xu, Y. Wang, B. Yang, and L. Hu, "Designing textile architectures for high energy-efficiency human body sweat-and cooling-management," *Advanced Fiber Materials*, vol. 1, no. 1, pp. 61–70, 2019.

8. M. G. Abebe, G. Rosolen, J. Odent, J.-M. Raquez, and B. Maes, "A dynamic passive thermoregulation fabric using metallic microparticles," *Nanoscale*, vol. 14, no. 4, pp. 1421–1431, 2022.

9. M. G. Garzon Altamirano, M. G. Abebe, N. Hergué, J. Lejeune, A. Cayla, C. Campagne, B. Maes, E. Devaux, J. Odent, and J.-M. Raquez, "Environmentally responsive hydrogel composites for dynamic body thermoregulation," *Soft Matter*, vol. 19, no. 13, pp. 2360–2369, 2023.

10. M. G. Garzon Altamirano, M. G. Abebe, J. Lejeune, A. Cayla, B. Maes, J. Odent, J.-M. Raquez, C. Campagne, and E. Devaux, "Design of silica coated polyamide fabrics for thermo-regulating textiles," *Journal of Applied Polymer*, e54004, 2023.

11. M. G. Abebe, E. Khousakoun, S. Desprez, J.-M. Raquez, and B. Maes, editors. P100_0202_ temperature-regulating textiles using switchable infrared reflectivity. Proceedings of the 19th World Textile Conference-Autex 2019, 2019.

12. M. G. Abebe, G. Rosolen, E. Khousakoun, J. Odent, J.-M. Raquez, S. Desprez, and B. Maes, "Dynamic thermal-regulating textiles with metallic fibers based on a switchable transmittance," *Physical Review Applied*, vol. 14, no. 4, p. 044030, 2020.

13. M. G. Abebe, A. De Corte, G. Rosolen, and B. Maes, "Janus-yarn fabric for dual- mode radiative heat management," *Physical Review Applied*, vol. 16, no. 5, p. 054013, 2021.

14. F. Pilate, A. Toncheva, P. Dubois, and J.-M. Raquez, "Shape-memory polymers for multiple applications in the materials world," *European Polymer Journal*, vol. 80, pp. 268–294, 2016.

15. J. I. Larruquert, A. P. Pérez-Marín, S. García-Cortés, L. Rodríguez-de Marcos, J. A. Aznárez, and J. A. Méndez, "Self-consistent optical constants of

SiC thin films," *Journal of the Optical Society of America A*, vol. 28, no. 11, pp. 2340–2345, 2011.

16. Y. Cengel and T. M. Heat, *A practical approach*. New York, NY: McGraw-Hill, 2003.

17. O. Takayama and M. Cada, "Two-dimensional metallo-dielectric photonic crystals embedded in anodic porous alumina for optical wavelengths," *Applied Physics Letters*, vol. 85, no. 8, pp. 1311–1313, 2004.

18. H. Aly, M. Ismaeel, and E. Abdel-Rahman, "Comparative study of the one dimensional dielectric and metallic photonic crystals," *Optics and Photonics Journal*, vol. 2, no. 2, pp. 105–112, 2012.

19. K. Yetisen, H. Qu, A. Manbachi, H. Butt, M. R. Dokmeci, J. P. Hinestroza, M. Skorobogatiy, A. Khademhosseini, and S. H. Yun, "Nanotechnology in textiles," *ACS Nano*, vol. 10, no. 3, pp. 3042–3068, 2016.

20. J. D. McCann, *Build the Perfect Survival Kit*. Penguin, 2013.

21. P.-C. Hsu, C. Liu, A. Y. Song, Z. Zhang, Y. Peng, J. Xie, K. Liu, C.-L. Wu, P. B. Catrysse, L. Cai, et al., "A dual-mode textile for human body radiative heating and cooling," *Science Advances*, vol. 3, no. 11, p. e1700895, 2017.

22. L. Cai, A. Y. Song, P. Wu, P.-C. Hsu, Y. Peng, J. Chen, C. Liu, P. B. Catrysse, Y. Liu, A. Yang, et al., "Warming up human body by nanoporous metallized polyethylene textile," *Nature Communications*, vol. 8, no. 1, pp. 1–8, 2017.

Index

Pages in *italics* refer to figures.